Powerline

Also by Paul Wellstone

The Conscience of a Liberal:
  Reclaiming the Compassionate Agenda

How the Rural Poor Got Power:
  Narrative of a Grass-Roots Organizer

POWERLINE

The First Battle of America's Energy War

Paul Wellstone and Barry M. Casper

Foreword by Senator Tom Harkin

University of Minnesota Press
Minneapolis • London

To Nancy and Sheila

Published by the University of Minnesota Press
111 Third Avenue South, Suite 290
Minneapolis, MN 55401-2520
http://www.upress.umn.edu

ISBN 0-8166-4384-9

Library of Congress Cataloging-in-Publication Data

Wellstone, Paul David.
    Powerline : the first battle of America's energy war / Paul
Wellstone and Barry M. Casper ; foreword by Senator Tom Harkin.—
1st ed.
      p.   cm.
   ISBN 0-8166-4384-9 (acid-free)
   1. Energy policy—Minnesota 2. Energy policy—Citizen participation. 3. Farmers—Minnesota—Political activity. 4. Power
resources—Minnesota. I. Casper, Barry M. II. Title.
   HD9502.U53 M68 2003
   333.793'2—dc21

                                    2003011745

Printed in the United States of America on acid-free paper

The University of Minnesota is an equal-opportunity educator and
employer.

Contents

Foreword   *Senator Tom Harkin*   vii

Acknowledgments   xi

1. Prelude to a Protest   3
2. The Powerline Is Coming   25
3. The Public Participates: The State Decides   55
   Profile: Jim Nelson   128
4. On the Brink of Violence   135
   Profile: Virgil Fuchs   149
   Profile: Ben Herickoff   154
5. A Populist Governor   165
   Profile: John and Alice Tripp   192
6. Confrontation on the Prairie   201
   Profile: Math and Gloria Woida   227
   Profile: George Crocker   234
7. Who Sacrifices, Who Benefits, and Who Decides?   241
   Profile: Dick Hanson   258
   Profile: Patty Kakac   265
8. The Towers Go Up: The Towers Come Down   273
   Epilogue   311
   Glossary   313

Foreword
Senator Tom Harkin

A local rebellion occurred in several rural counties of west central Minnesota in the mid- to late 1970s over the decision to locate an electric powerline through good farmland there. The campaign of opposition by farmers and their supporters is now an established chapter in state history. Lore about the "bolt weevils" who toppled giant steel transmission towers at night has been passed on as a compelling image of rural people's resistance against power companies and state authorities. The events of the powerline struggle amount to a noteworthy episode of populist protest, and this book remains that story's most thorough telling.

The fact is, though, that today we probably read this book mostly because we know that one of the story's tellers, an activist college professor at the time, Paul Wellstone, eventually became a U.S. senator—a senator who would make unique and dramatic contributions to our country's history and political life in his far-too-brief career. After cowriting this book, Paul went on to participate in, teach about, and write on other Minnesota and national political struggles throughout the 1980s. Then, in 1990, he was elected to the U.S. Senate. He promptly carried the kind of fight and values described in this book straight to Washington, where he shook things up for twelve years and rocked the system right up until the terrible day that his life was tragically cut short.

As I've said before, Paul Wellstone was my best friend in the Senate. He truly was the soul of the Senate, and no one ever wore the title "Senator" better—or used it less. Staff and citizens alike called him Paul, and he wouldn't have had it any other way.

Paul was always the same person. As political scientist and teacher, as community organizer, or as representative of the citizens of Minnesota, his method was based in direct and personal connection to people. He blurred the distinction between his life projects. He organized and agitated in the classroom. He encouraged and gave strategic advice to the subjects of his social science interviews. He taught from the floor of the U.S. Senate.

Paul was deeply committed to his political principles. But he was also a keen listener and respected the motivations of the people he spoke with, including those who differed from him. People felt this, and it helped him excel at all three of his professional callings.

Paul worked with a true partner on this book, Barry "Mike" Casper, a physicist colleague and friend at Carleton College who was every bit Paul's equal when it came to political passion. Mike's grasp of technology and energy policy helped frame what farmers saw as a fight rooted in their moral obligation to the soil and the future. At a key moment during this period, Mike joined one of the protest leaders, Alice Tripp, on a ticket to try to deny the Democratic–Farmer Labor nomination to incumbent governor Rudy Perpich. They gained a surprising 17 percent of the primary vote. Mike became one of Paul's first staffers in Washington and drafted an alternative national energy policy bill when Paul joined the Senate Energy Committee.

Powerline is about power, both electrical and political, and about justice. It is also a study of political process—not the way textbooks say it works, but the way it really happened in Minnesota. Power companies, cooperatives that were supposed to be run by and for their rural customers, used eminent domain and plotted their line according to a dry, numerical "avoidance rating" system. The system assigned the lowest avoidance value to farms. How out of touch could they be? No matter: farmers who resisted the companies' plans found the process rigged and their state government tilted against them. Local law enforcement was reluctant to use force against those engaged in peaceful civil disobedience, but the bureaucratic hearings would not go farmers' way. They didn't win their fight, even though more than 60 percent of Minnesotans supported their cause. They were outmatched by the power companies' lawyers and technical experts. In the end, state government and the courts took the companies' side.

The story's outcome is sadly familiar, but its telling is often inspiring. Many of the accounts here are firsthand, and some are humorous. Hog manure was put to tactical use on more than one occasion. Bullets and deadly dangerous anhydrous ammonia were also used, however, and both sides of this fight considered the stakes to be very high. The interviews that make this book possible could only have happened after the authors gained deep trust from farmers, some of whom at minimum had knowledge of serious illegal acts.

The farmers' struggle did affect energy policy in the state, though not immediately. Their protests raised questions about a paradigm of remote, unaccountable decision making that led to huge, centralized power production and lines that disrupted the landscape. These questions are around today, and Minnesota, like my home state of Iowa, has since done much to involve communities in energy planning, as well as to diversify sources to include more renewables such as wind and biofuels.

The powerline fight also prefigured later political events. It may even have helped rekindle a Midwest tradition of populist prairiefire that connected urban and rural dissatisfaction. Paul and Mike devoted much of their time and talents in the 1980s to the family farm movement, even while they were deeply immersed in peace activism, organizing the poor, and working on other issues that mobilized the Twin Cities' large progressive community. Paul helped bring students and labor union members from metropolitan areas to support striking P-9 meatpackers in Austin, Minnesota. By 1988, he was Minnesota chair of Jesse Jackson's run for the Democratic presidential nomination, and the network he developed in that campaign became the foundation for the ground game of his famous outsider race to the Senate two years later.

Senator Paul Wellstone didn't emerge suddenly or magically from the powerline fight in west central Minnesota. But he didn't come from nowhere, either. His approach to politics resonates clearly as he helps to tell the story in this book. That makes it a story worth remembering.

Acknowledgments

Our special thanks go to Monte Tarbox. He helped to plan the research, did many interviews and contributed numerous valuable ideas. One of the most enjoyable aspects of this project was the opportunity to work with him.

Many friends helped immensely with valuable criticisms of early drafts of the manuscript. Among them were Harry Boyte, Jonathan Casper, Nancy Casper, Rosemary Chalk, Bruce Hanson, Frances Fox Piven, Joyce Scholz, Karl Scholz, Seymour Schuster, Norman Vig, Carl Weiner, Stan Wenocur, and Sheila Wellstone.

We are especially grateful to Andy Driscoll for providing us with the transcripts of the interviews he conducted in preparation for his powerful television documentary about the CU project, "Powerplay."

Virgil Fuchs was in this protest from the beginning and he always brought his tape recorder along. His tapes provide a verbatim record of the early meetings on routing the powerline, the organization of the protest, the confrontations and the rallies, and the later meetings between the protestors and state and co-op officials. Access to these tapes was an invaluable aid to this research, for which we express our deep gratitude.

Without the help of Marilyn Schuster, Nansi Price, Sigrid Gronning, and Janet Runkel, we would never have been able to transcribe and use the hundreds of hours of interviews we had taped. Janet Runkel's superb help in typing draft after draft of the manuscript made what could have been an arduous task into a pleasant experience.

Thanks also to Julie Copeland of ECOL at the Minneapolis Public Library, who made it easy for us to study the EQC corridor and route hearings on this powerline. The Minnesota Energy Agency Library and the Minnesota Historical Society provided other valuable records on microfiche and microfilm.

Many talented people created songs that commemorate and chronicle this protest. We are indebted to Nancy Abrams, Larry Long, Henrietta McCrory, Russell Packard, and Minneapolis' Unity Theatre for allowing us to include excerpts from their songs in our book.

The Youth Project, Carleton College, the Ford Foundation, and Pro Bono Publico Foundation provided us with financial support at various stages of the project. Their assistance made this book possible. Carl Weiner was a friend indeed when he arranged for the small but critical Pro Bono Publico grant that enabled us to do the final round of interviewing and to afford secretarial help in preparing successive drafts of the manuscript.

One of us (B.M.C.) would like to thank several institutions for their kind hospitality. Immediately prior to the powerline dispute, research fellowships from the Rockerfeller Foundation and the National Science Foundation at Harvard University's Program for Science and International Affairs, the Brookings Institution, and the University of Minnesota's School of Public Affairs provided the opportunity for many ideas about energy policy, the science court, and related technology policy concepts to crystallize. MIT's Program in Science, Technology and Society has been a stimulating place to be as the book received its finishing touches.

J. W. Bradley, Kate Bradley, Charles "Boomer" Winfrey, and Maureen O'Connell from the mountains of east Tennessee, leaders of the courageous organization SOCM—Save Our Cumberland Mountains—have been an inspiration to one of us (P.D.W.). This book is partial repayment for all they taught.

Many local, state, and federal government and power cooperative officials have generously shared with us their recollections, their reflections, and their ideas about the CU powerline controversy, and they have made many documents and other records available to us during the course of this study. We very much appreciate the help we have received from them.

The family farms across western Minnesota have been our second home during the past three years. We owe our thanks to all the farmers who have provided information and who have expressed to us not only their ideas, but also their deepest feelings about their struggle. In the course of writing this book, we have come to understand and sympathize with their position. When Alice Tripp, a farmer and protest leader, challenged Governor Rudy Perpich in the Democratic-Farmer-Labor primary in September 1978, one of us (B.M.C.) became her running-mate as candidate for lieutenant governor.

During this study, we have spent literally hundreds of hours in interviews and in trying to understand the perspective of each individual we have contacted. To all who contributed, we extend our thanks and our pledge that we have endeavored throughout this book to represent their positions honestly and fairly.

Perhaps our chief regret as we finish writing the book is that some individuals who played important parts in the story have been inadvertently left out or have not received the credit they deserve. To them we offer our sincere apology.

We owe a special debt to Leone Stein, director of the University of Massachusetts Press. We appreciate her enthusiastic support for and sympathetic understanding of what we would like to accomplish with this work.

We would also like to express our appreciation to everyone else at the University of Massachusetts Press who contributed so much to this project, especially Pam Campbell for ably setting the type, Deborah Robson for a superb job of copyediting, Mary Mendell for creating such an attractive book and for treating us so kindly, and Ralph Kaplan who has patiently and professionally helped to build support for our book.

Powerline

They're stripmining coal in North Dakota

Generating power there

But they're stringing their lines across Minnesota

Hoping that we just don't care.

—from "Minnesota Line" by Nancy Abrams

1   Prelude to a Protest

To most Americans, the high voltage powerlines that crisscross our countryside are just a fact of life—links in an energy network whose existence is essential to our modern way of living. To many Minnesota farmers, however, one powerline has become a powerful symbol—a symbol of America's willingness to sacrifice its rural citizens to feed a gluttonous hunger for energy.

In 1973, two cooperative utilities, Cooperative Power Association of Edina, Minnesota, and United Power Association of Elk River, Minnesota, announced plans to build a large electricity generating plant on the site of a lignite coal strip mine near Bismarck, North Dakota. The electricity generated by the plant would be transported eastward 430 miles to the outskirts of the Twin Cities of Minneapolis and St. Paul in Minnesota. The entire undertaking, known as the CU project and mapped in figure 1, would be financed by low-interest loans provided through the Rural Electrification Administration (REA) in Washington. It would be the largest single project in the history of the REA.

The CU powerline and the protest it precipitated are unique. The line carries direct current (DC), not the usual alternating current (AC), and the eight hundred thousand volts between its conductors make it and a similar line on the west coast the highest-voltage DC powerlines in the United States.

The protest by Minnesota farmers whose land was crossed by the line began quietly in 1974, as legal channels of opposition were tested, built to a crescendo in 1978, when state troopers were used to keep angry farmers from chasing surveying and construction crews from the farmers' fields, and entered an unexpected and unprecedented phase of guerrilla warfare after construction was completed.

The protest preoccupied a governor, Rudy Perpich, from the time he came to office in January 1977 till the time he left two years later. As he said in retrospect, "It took more time, more effort than anything else I did. Probably more time and effort than the next five things, you know. Get up in the morning and just wonder, just what's going to happen that day."

In early 1978, when hundreds of farmers rose up in anger and went into the fields to stop surveying and construction, Perpich moved over two hundred state troopers, nearly half the Minnesota Highway Patrol, into tiny Pope County in west-central Minnesota to allow the line to be built. For months the attention of Minnesotans was riveted on dramatic confrontations in the fields. The national media came to cover the weaponry and tactics of a spectacular new front in the energy war—as angry farmers deploying tractors, manure spreaders, and ammonia sprayers confronted lines of troopers armed with guns and mace in the deep snow and bitter cold of a Minnesota winter.

But what is really significant about this protest is what happened before the media came on the scene and what happened after they left. The story does not begin with the protests in the fields and it does not end with the completion of the line. It begins long before, with the proposal to build the line and the processes by which state agencies approved the proposal and selected the route the line would follow. From one perspective, the project is an integral part of a farsighted national energy policy, of which increasing use of western coal is an essential element. By some accounts the procedures used to approve the proposal and select the route are a model of democratic practice. But those called upon to make the principal sacrifices for the powerline, the farmers, were not convinced of the wisdom of the enterprise and hence of the necessity for their sacrifice. At the same time they were alienated by the procedures employed to approve the project and to take away their land.

Perhaps more significant is what happened after their protests failed to stop the line. When construction was completed, the governor and the media declared the protest over. The democratic process had worked its will; the farmers had participated fully, but had lost; the "will of the majority" should be accepted. But a strange thing happened. The farmers did not stop protesting. A kind of guerrilla warfare broke out on the western plains and 430 miles of powerline proved difficult to defend. At night, towers were attacked by "bolt weevils" and fell to the ground; the land was littered with glass as an epidemic of "insulator disease" broke out; and high-powered rifle bullets (or was it "wire-worms"?) splayed open the conducting wire. Local sentiment was with the farmers and law officers made no arrests. Even with the most modern technology, including high-speed helicopters, utility security personnel and state police were frustrated.

What happened in Minnesota and why it happened may have a significance for our nation's energy policy far beyond the questions involved in building just one powerline. A potent new force —rural Americans—has something to say about what that energy policy should be and they may have discovered a source of power that will make all America listen.

We will consider the farmers' perspective later. First it is important to understand the conception of the cu project.

Figure 1. The cu powerline, from a power plant at the Coal Creek lignite mine, near Underwood, North Dakota to the Dickinson substation, just outside the Twin Cities in Minnesota.

Coal is the Answer

The CU project epitomizes the direction of U.S. energy policy in the 1970s and early 1980s. Central to that policy is increasing reliance on electricity—and the thrust in electricity is toward large, central generating facilities and more use of coal.

This direction is natural, given the institutions that make our energy policy. It is the utilities that bring energy, be it electricity, oil, or natural gas, to our homes. The planning for and construction of energy facilities are effectively in their hands.

To be sure, the utilities are subject to government regulation. But, as in the case of the CU project, that regulation tends to be only weakly coupled to the decision-making process. The utilities make the plans; the regulatory agencies react to those plans, usually after they are well along. The utilities have access to information and expertise; the regulatory agencies frequently have to rely on information provided by the utilities, and tend to be understaffed and overloaded. This leads to a special relationship between the regulators and the regulatees, with shared assumptions about what is appropriate and what is possible.

One especially significant aspect of policy-making for electricity is the relationship among the utilities themselves. They are tied together in power pools and they contract with each other to buy and sell electric power and to "wheel" one another's electricity along their transmission lines. For example, UPA and CPA are both members of the Mid-continent Area Power Pool (MAPP), an association of twenty-eight utilities serving seven midwestern states. Through MAPP, the managers of the utilities coordinate the planning of new generating facilities and transmission lines. It is within the context of the highly integrated network of MAPP that the CU project was planned.

In planning to tap western coal, the CU proposal fit a principal thrust of U.S. energy policy. With scarce and uncertain supplies of petroleum needed for other purposes, with nuclear power out of favor, coal is the obvious alternative for centralized generation of electricity. Increased use of coal has been a centerpiece of the energy strategy of three successive American presidents. This is what they said in energy messages to Congress while the CU powerline was being planned and built:

Richard Nixon (April 18, 1973): More than half the world's total reserves of coal are located within the United States. This resource alone would be enough to provide for our energy needs for well over a century.... I urge that highest national priority be given to expanded development and utilization of our coal resources. Present and potential users who are able to choose among energy sources should consider the national interest as they make their choice. Each decision against coal increases petroleum or gas consumption, compromising our national self-sufficiency and raising the cost of meeting our energy needs.

Gerald Ford (February 26, 1976): Coal is the most abundant energy resource available in the United States, yet production is at the same level as in the 1920s and accounts for only about 17 percent of the Nation's energy consumption. Coal must be used increasingly as an alternative to scarce, expensive, or insecure oil and natural gas supplies. We must act to remove unnecessary constraints on coal so production can grow from the 1975 level of 640 million tons to over 1 billion tons by 1985 in order to help achieve energy independence.

Jimmy Carter (April 20, 1977): Although coal now provides only 18 percent of our total energy needs, it makes up 90 percent of our energy reserves. Its production and use do create environment difficulties, but I believe that we can cope with them through strict stripmining and clean air standards. To increase the use of coal by 400 million tons or about 65 percent—we now use about 600 million tons—in industry and utilities by 1985, I propose a sliding scale tax, starting in 1979, on large industrial users of oil and natural gas.... I will also submit proposals for expanded research and development in coal. We need to find better ways to mine it safely and to burn it cleanly and to use it to produce other clean energy sources like liquefied and gasified coal. We have already spent billions of dollars on research and development on nuclear power, but very little on coal. Investments here can pay rich dividends.

As one essential element in implementing this policy, national energy planners looked to a dramatic increase in the stripmining of western coal. For example, one projection of expanded coal production in 1974, in the well-known *Project Independence* report, foresaw annual stripmining production in North Dakota, South Dakota, Montana, and Wyoming alone shooting up from a 1973 level of 32 million tons per year to 298 million tons per year in 1985. Another projection, by the Federal Energy Administration in 1976, is shown in figure 2. Clearly an enormous expansion in stripmining western coal was anticipated at the time the CU project was initiated.

Origins of the REA

On May 11, 1935, by executive order of President Franklin D. Roosevelt, the United States government undertook an historic mission—the Rural Electrification Administration was estab-

Figure 2. Where will new coal production come from? Source: Federal Energy Administration, *National Energy Outlook* (February 1976).

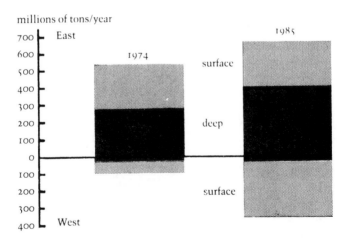

lished to bring electricity from central generating stations to light up rural America. At that time only about 11 percent of America's farms had electricity. The technology for transmitting electric power long distances had been available for decades, but commercial electric companies were asking far more than the average farmer could afford.

The original idea of REA was to provide financial incentive to the commercial electric companies by offering them low-interest loans to expand into rural areas. But the response was disappointing. According to REA's first administrator, Morris L. Cooke, "before December [1935] . . . it became apparent that the industry was not going to use even a substantial portion of the funds available for rural electrification. . . ." But farmers wanted that electricity and many were ready to take on the job themselves. In Cooke's words, "farm organizations of a cooperative character forged to the front as the principal borrowers under the REA program."

Congress moved quickly to support the cooperative approach to rural electrification. Under the sponsorship of Senator George Norris of Nebraska and Representative Sam Rayburn of Texas, Congress passed the Rural Electrification Act of 1936, which established the REA as a lending agency and gave clear preference to nonprofit associations. This was all the farmers needed. With guidance and support from REA in Washington, the rural electric cooperative (known for short as an *REC* or distribution co-op) became a feature of rural America. Since the co-ops were self-regulated by their memberships through member-elected boards, many states passed laws exempting them from the regulations normally applied to electric companies.

The result was, by all accounts, a stirring story of success. With the assistance of REA, electric power in a very short time transformed the character of life in rural America. Despite a slowdown during World War II, 78 percent of farms had central station electric service by 1949. REA, by then a part of the agriculture department, provided 2 percent loans to co-ops until 1973, when over 99 percent of America's farms had electricity. Nearly a thousand co-ops were operating and REA loans had grown to about $400 million per year.

In 1973, after Richard Nixon's reelection, REA's loan-granting authority was changed. The interest rate on direct REA loans was

increased to 5 percent and at the same time REA was granted unlimited authority to guarantee loans from the newly established
Federal Financing Bank (FFB) of the treasury department. Since
that time, the size of REA's loan-granting activity has jumped
enormously. By 1979, REA was granting about $1 billion per year
direct (mostly 5 percent) loans and had about $17 billion in FFB
loans (from 7 to 9 percent) outstanding.

The original concept of REA envisioned the primary function of
the co-ops as the distribution of electric power. It was assumed
that private electric companies would continue to operate most
of the central generating facilities. But in some areas the private
companies were less than fully cooperative with REA. Many distribution co-ops then combined in federations to provide their
own supply of electricity. A federation, known as a generation
and transmission cooperative or G & T, would purchase power
from private utilities, from federal agencies such as TVA or the
Bureau of Reclamation, and from municipal utilities; in some
cases it would borrow REA money to build its own generating stations. Today there are over forty generating and transmission co-
ops providing power to the distribution co-ops.

Cooperative Power Association (CPA), with offices in Edina, and
United Power Association (UPA) of Elk River are the Minnesota
generation and transmission cooperatives. CPA is a federation of
nineteen distribution co-ops, serving about one hundred and
thirty thousand member-customers in western and southern
Minnesota. UPA is a federation of fifteen distribution co-ops serving about one hundred and sixty thousand member-customers,
mostly in northeastern Minnesota. Prior to the CU project, CPA
was essentially a paper organization with about a dozen staff
members. It had no generating capacity of its own; instead it held
contracts to purchase power, primarily from two sources: 110
megawatts of hydropower from the U.S. Bureau of Reclamation
and 170 megawatts from a Wisconsin G & T, Dairyland Power
Cooperative. UPA had its own generating plant, the 150-megawatt
Stanton facility in North Dakota, and it purchased power from
several other members of MAPP, most notably Northern States
Power (NSP), the giant of Minnesota's investor-owned utilities.

From Competition to Partnership

During the Nixon years, the relationship between the generation and transmission co-ops and the investor-owned utilities (IOU's) underwent a significant change. From the beginning of REA, there had been tensions between the utilities and the co-ops. These tensions reached a head in the 1960s as suburbia penetrated previously rural areas, resulting in competition for customers. In the 1960s the investor-owned utilities complained bitterly about the unfair advantages the co-ops had—notably, access to a ready supply of low-interest money from the federal government and much less regulation by state governments. One no longer hears such complaints today. The investor-owned utilities have in effect formed partnerships with the generation and transmission co-ops and now benefit from their advantages.

Closer ties between the utilities and the cooperatives came about naturally. Huge central generating stations have to be linked in a grid, so that a shut-down in any one of them can be compensated for by power from others. If periods of peak demand by customers of one utility can be met by the generating capacity of another, the overall system is more efficient and cost-effective. Once a grid is established, much coordination among utilities is required. Hence the genesis of regional power pools like MAPP.

These partnerships between utilities and cooperatives established in the name of reliability, efficiency, and cost-effectiveness, have other consequences. Federal monies available to the cooperatives so they may bring electricity to rural America are now accessible to investor-owned utilities. This may occur directly, as, for example, in the case of an agreement between UPA, NSP, and other utilities to build jointly a nuclear power plant in Wisconsin. It may occur indirectly if electricity from a generating plant built in North Dakota by UPA and CPA is in part funneled through the grid into the homes and businesses of NSP customers. It's hard to keep that electricity down on the farm after it's seen NSP.

Pressures to bring REA money into the grid are institutionalized by agreements such as the one UPA and CPA had to sign in order to join MAPP. The MAPP agreement obligates each member utility to provide for its own power requirements. By signing, UPA and CPA had come a long way from the original, limited REA mission of using government money to insure that transmission lines would reach rural America. They had become full partners with the investor-owned utilities in an interlocking network that reaches all America and formed a direct link between that network and the federal treasury.

The REA and North Dakota Lignite

One more bit of history is necessary to set the stage for the CU project. That is the pioneering effort by the REA in the North Dakota coal fields. Coal lies beneath the entire western half of North Dakota, but this coal is unlike that found anywhere else in the country. It is called lignite and it does not burn very well. Due to its high moisture content, one ton of lignite has roughly half the energy content of the bituminous and subbituminous coal found elsewhere, including under the fertile fields of Montana and Wyoming. As a consequence, unlike Montana and Wyoming coal, it is uneconomical to ship lignite long distances by train to generating plants.

Back in the 1950s, using lignite to make electricity looked pretty risky. Still, North Dakota is half lignite, so REA took a gamble. Co-ops in Minnesota and the Dakotas had little generating capacity of their own. Instead, they relied heavily on hydropower from a series of Bureau of Reclamation dams on the Missouri River. As they projected their exponential growth in demand, it appeared that additional sources of supply would be needed by the late 1960s. In 1959, with backing from the REA office in Washington, the co-ops decided to try lignite. A new G &T, Basin Electric, was created and its first project was to build a generator right in the lignite field—the two-hundred-megawatt Leland Olds Plant at Stanton, North Dakota, on the Missouri River north of Bismark. In the words of an REA official who worked closely with the project, Frank Bennett, "It was sort of experimental, but it worked. . . . We put our toe in the water in 1960 and it worked great." An REA loan was granted in 1961, construction began in 1962, and the plant came on line in 1966.

The Olds plant was such a success that it was soon followed by two similar facilities financed by REA. One was also located at

Stanton—UPA was created to build that plant in 1967. The other was the two-hundred-megawatt Center facility. In the early 1970s, all three plants were producing electricity at a cost of only seven to eight mills per kilowatt-hour, which placed them all among the ten most economical plants in the country. In fact, the Center plant was at the very top of the list. REA was proud of its accomplishment; it had pioneered the use of a plentiful energy resource, lignite coal. As Bennett recalls, "By 1973, we had plants with five to six years' experience, working wonderfully well." REA was ready to move up to even bigger lignite plants.

This is the background to the CU project. In 1972, UPA and CPA approached REA about the possibility of building a new generating facility. According to Theophile V. ("Ted") Lennick, general manager of CPA,

> Way back, I suppose in the early 70s, probably in 1972, CPA looked at its load requirements for the future and determined that we needed something like about 450 to 500 megawatts to take care of our requirements through the period 1980 to '81, probably into 1982 . . . at the same time United Power was looking at their needs and discovered they too needed something in the neighborhood of 400 megawatts, so we got together with them and compared notes. . . .

The CU Project is Conceived

In June 1972, Frank Bennett flew out to Minneapolis for a planning meeting with UPA and CPA officials. A number of other co-ops, including Dairyland from Wisconsin, the nation's largest G & T, Minnkota from North Dakota, and some Iowa co-ops were represented at that meeting. They agreed to explore the possibility of a joint generating facility in the North Dakota lignite fields, and hired an engineering consulting firm, Burns and Roe, to do a feasibility study. On the basis of that study, Dairyland dropped out ("too far from the lignite fields," according to Bennett); so did Minnkota and the Iowa co-ops. That left UPA and CPA, who commissioned another engineering firm, Burns and McDonnell, to do a study comparing the options of a North Dakota lignite plant and of a Minnesota plant using Montana coal shipped in by train.

The REA Washington office clearly favored the lignite option. In Bennett's words, "We had three of the top ten generating plants in the United States . . . with that record of success and that record of accomplishment, how could you go wrong? . . . It was the REA's preference to go lignite fields rather than Minnesota plant. . . . It was right as rain to do this . . . we realized we were coming to an energy crunch and this would be a good thing."

With this clear preference from the men with the money in Washington, it is not surprising that a large lignite facility in North Dakota was chosen. Burns and McDonnell wrote a report that said it would be more cost-effective than a Minnesota plant.

UPA and CPA thus chose to build a mine-mouth plant in North Dakota and REA was pleased to fund it. Frank Bennett recalls, "At that point in time, lignite coal fitted best into our national energy policy and to exploit the lignite coal reserves in North Dakota seemed like the best engineering and economic solution to our energy-load problem."

At this point, another important actor, North American Coal Company, did something that might have thrown a monkey wrench into these well-laid plans. North American, the nation's largest independent producer of coal, had been a partner in the previous North Dakota lignite projects. In 1973, North American was approached formally by UPA and CPA. It replied with a letter of intent that indicated it would supply lignite to the new plant at a price of nineteen cents per million Btu. With that understanding, in February 1974 REA made a conditional loan to UPA and CPA, based on satisfactory development of the coal lease.

But North American unexpectedly changed its tune. Frank Bennett tells the story this way: "All of a sudden North American Coal Company got difficult to deal with. . . ." They said, "All past bets are off," and demanded unprecedented terms for their participation in the project. They would have to be guaranteed a profit, beginning with a price of one dollar per million Btu, over four times what they had initially offered and more than twice the price of Montana coal with delivery charges included. In addition, the co-ops would have to pay all coal mine development expenses.

REA officials were taken aback by this turn of events, but undaunted. They decided to continue with the project—on North

American's terms. Falkirk Mining Company, a wholly-owned subsidiary of North American, was created in August 1974 for the sole purpose of producing lignite for the eleven-hundred-megawatt Coal Creek generating station, to be built on the Missouri River near Underwood, North Dakota. Falkirk and the co-ops signed a coal agreement, the text of which has not been made public. It is known, however, that Falkirk is to provide up to six million tons of lignite annually for thirty-five years and REA's Frank Bennett has stated, "It's the first one I've been involved in where we financed the whole mine."

The Strip mine, the Power plant and the Powerline

With the signing of the coal agreement in October 1974, the final hurdle was cleared and the CU project was underway. The project consists of three principal parts—the Falkirk mine, the Coal Creek generating station, and the CU powerline. Each part is a remarkable example of innovative technology.

When the mining operation reaches its peak production rate of 5.6 million tons per year, it will be one of the largest strip mines in the United States and the leading producer of coal in North Dakota. It covers over twenty-five square miles and employs two of the largest drag-lines ever constructed. These mammoth machines roam the mine site, stripping as much as 135 feet of overburden to reach the lignite. Their size has been graphically described by Falkirk president Robert Murray: "One of these machines could be positioned between two football fields placed end-to-end, and it would be able to lift 152 tons from well into the end zone of one football field and place it in the end zone of the second football field over 218 yards away." The lignite uncovered by the drag-line is then crushed and shipped by conveyor belt to the generating station.

The 1100-megawatt Coal Creek generating station consists of two 550-megawatt units in huge buildings that tower over the North Dakota prairie. The alternating current produced by the generator is immediately fed into an AC-DC conversion station, from which direct current emerges.

This direct current is transmitted from Underwood 430 miles eastward by a +/−400 kilovolt powerline to a DC–AC conversion

station near Delano, Minnesota, just outside the Twin Cities. The
powerline has two pairs of wires and "+/−400 kilovolts" means
that the voltage difference between one of the pairs of wires and
the ground is +400,000 volts and the voltage difference between
the other pair of wires and the ground is −400,000 volts. The voltage between the two pairs of wires is 800,000 volts.

The powerline wires, about 1.5 inches in diameter, are strung
on giant steel towers, averaging 180 feet high, with about four
towers per mile. This line is one of the first of its kind ever constructed. The Bonneville Power Administration's Pacific Intertie
from northern Oregon down through Nevada and into the Los
Angeles area is the only comparable powerline in the United
States. It is also +/−400kv DC and was completed in 1970.

There are economic and technological reasons why high voltage direct current (HVDC) lines were not employed in the past and
why this technology was chosen for the CU project. One advantage of HVDC is that less energy is lost per mile than with an AC
line. A disadvantage of HVDC is that energy is lost in converting
from AC to DC at the generating plant and back to AC again for delivery to customers. Recent advances in conversion technology,
however, have reduced these losses to the point where HVDC
transmission is more cost-effective if the distance the power is to
be transported is sufficiently long.

This economic argument is the one usually cited for the choice
of HVDC for the CU line. But REA's Frank Bennett asserts there was
another, technical reason for choosing DC. "The basic reason for
going DC," he contends, "was to help the transmission grid." The
CU line is the single major link between the western part of the
MAPP grid in the Dakotas and the part in the Twin Cities area. If
that link were AC, serious instability problems could result, with
a fluctuation in North Dakota creating outages in Minneapolis
and vice versa. A DC line reduces such instabilities.

To determine a route for the powerline, the co-ops turned to experts. In 1973, UPA hired Commonwealth Associates, a Michigan
consulting firm, to propose a route and to assess the environmental impact of the line.

Commonwealth used a computer. A grid was laid over a map of
North Dakota and Minnesota, dividing the land into squares one
mile on a side. Each square was then assigned a number from zero

to six: the higher the number, the more important it was to avoid that area in routing the line. This information was fed into a computer and used to select a route that minimized adverse environmental impact. Of course, what a computer decides depends on the assumptions built into its decision-making process by people. The people from Commonwealth Associates decided what number to assign to each square. They assigned high numbers to airports and highways and wildlife areas, moderate numbers to woodlands, and zero to agricultural land. The resulting route was predictable; it avoided airports, highways, and wildlife areas, and cut diagonally across many farms.

How Technology Gets Deployed

In retrospect, it is easy to see how the CU project came about. In 1972, REA was ready to move up to a really large lignite generating plant; UPA, CPA, and the other MAPP utilities wanted the electricity; the just completed Bonneville Pacific Intertie showed that the HVDC technology was available.

This is typical of the way new technology is introduced. Planners, looking ahead, conceive projects that promote institutional objectives. Experts are hired to decide on the technical details and to write the required reports. Monies, public and/or private, are acquired from sympathetic sources. These monies are invested and the project is underway. Only then does word reach beyond the tight circle of planners and benefactors.

REA planners, UPA and CPA managers, executives of other MAPP utilities, and the management of North American Coal Company conceived the CU project. It fit the thrust of the national energy policy toward increased use of coal and especially promoted the special role one part of the REA bureaucracy had carved out for itself in support of that policy—encouraging the use of North Dakota lignite. It fit the objective of UPA and CPA managers to find electricity to meet their projected demand, especially since in the process the organizations they managed grew enormously in size and influence. It fit the objective of the executives of other MAPP utilities to strengthen the regional electric grid, especially since financing came from the federal government. It fit the objectives of the owners of North American Coal to make a profit, especially since the mine was built without cost to them.

Experts at consulting firms like Burns and McDonnell and Commonwealth Associates provided technical advice and produced reports that supported the lignite option, justified the use of HVDC, and assessed environmental impacts.

Falkirk Mining agreed to provide the coal; CPA and UPA agreed to build the power plant and the powerline; the REA agreed to guarantee the loans. In 1974, the CU project was underway.

One thing missing from the decision-making process was the voice of co-op members, ironic in view of the original REA conception of the co-ops as democratic, member-run organizations. To be sure, each distribution co-op had an elected representative sitting on the board of either UPA or CPA. But, as in the case of most organizations that hire professional managers to run their day-to-day operations, the managers had become the major policy makers at UPA and CPA. The *de facto* policy-making forums for the co-ops were not the UPA or CPA boards, but rather the councils of MAPP, where managers of UPA and CPA met with other utility managers. Furthermore, while each distribution co-op had its own board, board members tended to serve a long time and eventually to act without significant consultation of the membership. In this way individual co-op members were doubly insulated from the processes whereby important decisions, like that to build the powerline, were made. When concern developed over the powerline, members found they could not even gain access to key documents, such as the coal agreement.

About all the members were supposed to do was to pay off the REA loan, through their monthly utility bills. Here some simple arithmetic is revealing. By 1979, UPA and CPA had borrowed $1.25 billion dollars for the CU project at an average annual interest rate of about 8.5 percent. There are approximately 300,000 households served by UPA and CPA. To build the coal mine, power plant, and powerline, each UPA and CPA household had in effect taken out an involuntary loan of over $4000 ($1.25 billion divided by 300,000), to be paid off over 34 years at 8.5 percent.

Democratic Decision-making

The environmental movement of the late 1960s and early '70s brought pressure on government at all levels to become more sensitive to the environmental impacts of its programs and the programs it regulates. It also brought pressure for more "public participation" in government policy-making and regulatory processes. The Arab oil crisis of 1973 triggered deep apprehensions about our energy future and produced pressure on government to participate more fully in long-term planning for energy needs.

In Minnesota, as in many other states, these political pressures resulted in institutional reforms. In 1973 and 1974, the Minnesota legislature passed an Environmental Policy Act and a Power Plant Siting Act; separate laws established the Minnesota Environmental Quality Council and the Minnesota Energy Agency.

Environmentalists had pushed hard for the Environmental Policy Act, which requires that an environmental impact statement be prepared for any project for which "there is potential for significant effects resulting from any major governmental action or any major private action of more than local significance."

The Power Plant Siting Act, passed later in 1973, states, "The Legislature hereby declares it to be the policy of the state to site large electric power facilities [i.e., generating plants and transmission lines] in an orderly manner compatible with environmental preservation and the efficient use of resources."

As usual in the legislative process, bargains were struck and compromises made. While the environmentalists won on many principles incorporated in the Environmental Policy Act and the Power Plant Siting Act, they lost on the crucial issue of who would be responsible for implementing those principles. The environmentalists originally wanted an independent three-person council and staff, modeled after the Council on Environmental Quality in Washington, which oversees the National Environmental Policy Act. State agencies, however, opposed such oversight of their activities. With the support of then-Governor Wendell Anderson, the legislature established the Environmental Quality Council (EQC), the majority of whose members are the agency heads themselves. The EQC has twelve members, including the head of the State Planning Agency (who is chairman of the

EQC), the heads of the Pollution Control Agency, the Department of Natural Resources, the Department of Agriculture, the Department of Transportation, the Department of Health, and the Energy Agency, as well as a representative of the governor's office and four private citizens appointed by the governor.

In support of this concept of the EQC, as opposed to a body independent of the agencies, it was argued that a council that included agency heads could perform a dual function. In addition to assuring environmental sensitivity, such a council would provide the much needed, but hitherto lacking, function of coordinated, long-term, state-wide planning among the agencies.

Another important institutional reform was the creation of the Minnesota Energy Agency by the legislature in May 1974. It so happened that Minnesotans had scarcely been inconvenienced while the populations of both coasts suffered severe gasoline shortages during the Arab oil embargo. Nevertheless, for a state where the temperatures regularly drop below zero for weeks at a time, the message was stark and clear. As the legislation creating the new Energy Agency put it, ". . . energy planning, protection of environmental values, development of Minnesota energy sources, and conservation of energy require expanded authority and technical capability and a unified, coordinated response within state government."

Under the new laws, the Energy Agency and the EQC were to decide two crucial questions about the proposed powerline: *whether* it would be built and *where* it would go. The new procedures by which these questions are to be decided ostensibly represent a sharp departure from the practices previously employed.

Whether a powerline would be constructed had traditionally been decided by the utilities. They planned a powerline, then went to county zoning authorities to request eminent domain rights. County approval was almost automatic; under the law, there was only one test of whether eminent domain should be granted and that was whether the proposed project served a "public purpose." Around the turn of the century the legislature had decreed that any electric utility project passed that test.

Where a powerline would be routed was also decided, in practice, by the utilities. To be sure, county officials had the authority to set conditions when granting eminent domain, but generally they went along with the route the companies had mapped out. Once the utilities received the go-ahead from the counties (and often before, since approval was virtually assured), they would go to individual landowners and offer to buy the easements for the line. In the case of the CU project, the easement, or "right-of-way," is a strip of land 160 feet wide, running the length of the line. The utility then became the owner of the easement. In the case of farms, however, the usual arrangement was to allow the farmer to continue to use the land in return for keeping the areas around the bases of the towers cleared. If a landowner refused to sell, the utility could exercise its right of eminent domain and go to court to acquire the land through condemnation. All the court would decide was "just compensation." Use of the courts was rarely necessary, however, since most landowners knew there was little to be gained.

The new laws changed this process. The right of eminent domain still exists, but *whether* and *where* are now decided in different forums, according to different standards. The criterion for deciding whether a powerline is to be built has been expanded from "public purpose," automatically satisfied by utilities, to "public purpose" *plus* a determination by the Energy Agency director that the powerline is *needed.* Before building any large energy facility in Minnesota, a utility has to apply to the Energy Agency for a certificate of need. Within six months after receiving an application, the director must issue a decision.

If a certificate of need is granted, the question of *where* the line is to be routed is determined by the Environmental Quality Council. In preparation for this process, the EQC was directed by the Power Plant Siting Act to scan the state and develop an inventory of potential high voltage line corridors. Once a utility receives a certificate of need from the Energy Agency, it applies to the EQC.

Route selection by the ECQ then proceeds in two parts, each taking six months. First a thirty-kilometer-wide corridor is selected; then a one-kilometer-wide route is chosen within the corridor. Both the processes of corridor selection and route selection involve much public participation. Citizen advisory committees are appointed to advise the EQC in each selection. At least one public hearing is held in each county containing a proposed

corridor and then, when the route is to be selected, public hearings are held in each county along the route chosen. Meetings of
the EQC and the hearings are open to the public, with a complete
record kept and available for public inspection.

In parallel with these processes, the Department of Natural Resources submits an environmental assessment to the EQC. If the
EQC deems it necessary, it can then request a full-scale environmental impact statement.

All these institutions were created by the legislature in 1973
and 1974 in the name of environmental sensitivity, state-wide
long-term planning, and citizen participation. Any large energy
facility, such as the CU powerline, would be subject to questions
beyond simply, "Is it technically feasible?" and "Will it be profit-
able?" There would have to be acceptable answers to such questions as, "Is it needed?," "What will it do to our environment?,"
and "Does it fit our long-range energy plans?" These questions
would be raised in public forums with adequate opportunity for
all citizens to present their perspectives and concerns. The entire
process would be open, with decisions arrived at openly by accountable public officials.

That was the rationale for a new institutional framework. The
CU powerline was its first test. All these questions were asked.
Citizens did participate as prescribed. Yet when it was all over
farmers by the hundreds went into the fields to stop surveyors and
construction crews, brigades of state troopers were required to
guard completion of the project, and bands of farmers with widespread community support went out at night to bring the towers
down.

One view of these events was given by Governor Perpich, when
he spoke on state-wide television just before sending out the state
troopers:

> Our political system has been responsible and responsive. . . .
> All of government has bent over backwards to respond to the
> protesting farmers. We have gone the extra mile. In light of
> this, I can find no justification for the unlawful activities
> now taking place. The state Supreme Court has ruled that
> the line may be built. It is my job to uphold the law . . . and I
> will. State law enforcement officers will provide assistance

to local authorities as they have requested. The people working on the line will be protected. I ask you to join me in calling for an end to the harassment of workers, the interference with construction crews, and the destruction of equipment. My fellow Minnesotans, we need your support for our system of law.

This view was echoed by western Minnesota's largest and most respected newspaper, the *St. Cloud Times*, in an editorial when the towers began coming down:

The destroyers seek to overrule a democratic process and subject their will through violence. Support for the destroyers undermines the very democratic system that has permitted the legitimate protesters to openly express their displeasure with the official decision. The line was approved after a long series of public hearings and other steps in a complicated government process. The results of the process —certification of the need for the line and assignment of a route—reflected conflicting needs and desires of the overall public. The results were upheld after numerous appeals. Various changes in law followed the protests. Legal avenues for individuals unhappy about compensation and other matters still remain. . . . Democratic government processes must be open to every element in the society. They must be responsive to all elements. However, once the democratic process produces a decision, that decision must be upheld. No one, generally, is perfectly happy with a democratic decision, because it takes everyone's needs into consideration and produces results acceptable to the greatest number. . . . The extremists behind the tower destruction must be stopped. . . .

Was it simply a matter of the farmers' being unhappy because they didn't get their way? Were those who took down the towers rejecting a democratic decision that took everyone's needs into consideration? To what interests *was* the political system responsive? The farmers' perspective on these questions is quite different from those of Governor Perpich and the *St. Cloud Times*.

Now they come along with a good old plan

They want to survey my own land

>   *What can a poor man do?*

Wanna build a powerline from North Dakota

Straight through my farm here in Minnesota

>   *Give me them Pope County Blues!*

—from "Pope County Blues" by Larry Long

diagonally across the middle of the fields he and his father farmed. He was especially upset that he had not heard anything about a powerline until this meeting. He knew other Grant County farmers would feel the same way: "I talked to a neighbor, Gerald Bates. He sells federal crop insurance, so at that time he knew a lot more people in the area. Between the two of us, we started a kind of chain reaction thing of notifying everyone along the route."

Owen Heiberg is editor of the *Herman Review*. He and his wife Michele own and run the newspaper, a weekly tabloid of between sixteen and twenty pages, circulation 1150. He also contributed to the chain reaction. Heiberg remembers, "Jerry Kingrey from Cooperative Power Association dropped by the office to deliver a small notice of a public hearing that he wanted in the paper. I didn't know who he was then. I said, 'Fine.' And then I happened to glance at the notice. My initial reaction—we'd been here maybe one or two years—was, 'Gee, this looks like something pretty big. I wonder how people around here feel about it?' So I called the county administrator to get more information. I carried the twelve-line notice of a public information meeting along with a front page story and a map of the proposed route."

Heiberg's story, headlined HEARING ON PROPOSED TRANSMISSION LINE IS SET FOR JULY 10, appeared in the *Herman Review* on the fourth of July. A few days later Jerry Kingrey came by the newspaper office to obtain a copy of the paper for his company's records. Heiberg gave him one and recalls what happened next: "He saw the story and for a moment something happaned to his face. He is a very smooth person, but for a moment the smoothness left his face. He said, 'Where did you get that story?' I told him I called the county administrator. Then he got under control. But this was the first inkling I got that maybe everything isn't right with the powerline."

The farmers in Grant County did not need much convincing to attend the "information meeting." The courthouse in Elbow Lake was packed with seventy to eighty farmers who had just heard for the first time about "this powerline." The utilities, mandated by the Rural Electrification Administration to hold public information meetings, described what would happen. The line had to be built; they would build it the least bothersome way for farmers, and they would provide farmers with generous compensation.

2   The Powerline is Coming!

Farmers across west-central Minnesota consider Jim Nelson the founder of their protest movement. It is ironic that this young, softspoken farmer, then twenty-eight years old (some of the older farmers call him "our Jimmy"), should have initiated such a volatile and dramatic struggle. But he was the one who brought the news to farmers that one of the largest powerlines in the United States would soon be crossing their land.

It was late June 1974. Jim Nelson recalls: "I'm a township officer and the chairman of the town board got word to me that there was a meeting at the Grant County courthouse that I might be interested in. I never really heard how he found out about it. At that meeting the power company was telling the Grant County commissioners that they were building a line through the county and where they would build it. There wasn't too much the county commissioners could do about it one way or another, but they wanted to tell them about it. That is the way the presentation was made—this is what we're doing and we just thought you should know, even though there's nothing you can do about it." [To avoid confusion it should be noted that Nelson and the other farmers we interviewed used the terms "power companies" and "power co-ops" interchangeably. That is the practice we shall follow in the remainder of this book.]

At the commissioners' meeting Nelson discovered that the powerline was routed almost right over the top of his home: "They had a map there. It's kind of interesting the way the power company works. They always—in the early days—only had one of anything. I finally talked the commissioners into making the company give me a copy." Nelson was upset about the line going

They hadn't counted on the presence of one man in the audience—Verlyn Marth. Jim Nelson recalls: "That was the first time I ever saw him—I didn't know who he was before then—but Verlyn, he was sitting there and he said, 'Could I say a few words.' He started out pretty friendly, but then he got up in front and harangued them for about twenty or thirty minutes. . . . He was very effective. I've never heard anything like it."

Marth's own recollection is this: "There was a reluctance initially for people to get up and speak anything adverse. I'm not sure I was the first. But I got up with the basic theme I always do at the first meeting—the deliberate policy of trying to break down this aura of respectability they [the power companies] try to establish. The name of my presentation was, 'You can beat this line if you get together like you are doing here and these people are only here to take your land away from you.' That kind of broke the floodgates."

Marth did indeed strike home. According to Jim Nelson, Marth "really broke it open." Suddenly the power company representatives who called the "information meeting" were faced with hostile questions from a roomful of angry farmers. A protest movement had begun.

Verlyn Marth is an enigma, certainly to the power companies, but also to his neighbors in Grant County. He and his wife spend their summers on the family farm outside Herman. The rest of the time they live in California, where Marth previously worked in the aerospace industry for twenty-five years. He has been involved in several major utility fights in California. That is about all most people know of Verlyn Marth. He answers questions about his personal background by saying, "I just take care of my own."

Owen Heiberg describes Marth as "certainly the most articulate opponent. . . . Jim Nelson was the organizer, Verlyn was feeding in the ideas and rhetoric."

The Grant County farmers held their own meeting on July 18. Nearly a hundred gathered in the basement of the community building in Elbow Lake. They quickly agreed an organization was necessary to fight the powerline. They elected a board of directors, with Jim Nelson as president. The group passed a hat for start-up money and decided on a name, No Powerline (NPL). The *Herman*

Review reported, "The name was chosen for its direct communication of the group's purpose."

The largest direct current powerline in the United States, the most expensive project in REA's history, involving over $500 million in loans and loan guarantees (later the figure would rise to over $1.2 billion), had been a well-kept secret for two years. The farmers, members of CPA distribution co-ops, felt betrayed and were determined to stop the line. Owen Heiberg said: "I was surprised at how rapidly and strongly opposition to the line developed. I wasn't sure whether the line would be opposed or applauded, but I found out in a hell of a hurry."

The focus of NPL's attention was the county board. Under provisions of the 1973 Power Plant Siting Act, the state Environmental Quality Council now had jurisdiction over siting powerlines. However, there was a "savings clause" in the legislation. Projects already underway at the time the law was passed could apply to be exempted from its provisions. CPA and UPA had elected to exercise this privilege and seek approval in the way that had traditionally been almost automatic—from county authorities.

NPL organized farmers throughout Grant County. Before, only a handful of citizens attended county board meetings; now several hundred farmers regularly came to "lobby." They pressured the commissioners with many questions. Who decided the line was needed? Where would it go? Why wasn't farmland given the same accord as wildlife areas, highways, railroads, towns, and cities? They saw the powerline as an eye-sore in a beautiful land—the land they worked and the land they lived on. Verlyn Marth described it as an "aerial sewer."

Questions were raised about health and safety problems—electric shock, ozone, long-term biological effects. The line was routed to cut straight southeast through the middle of Grant County. As a consequence it would cross diagonally over many farmers' fields, rather than follow property and section lines. This would seriously disrupt such farming operations as irrigation and aerial spraying and seeding. Some wondered whether this would be the first of many powerlines to be built over their land. And over and over again the question arose—Why were these decisions made without asking us? Why weren't we considered?

NPL's first victory came on Monday, August 5, when the county board unanimously agreed to a ninety-day stay on any action concerning the line. NPL then persuaded the Grant County planning commission to recommend a new utility ordinance requiring a special-use permit for high voltage transmission lines. On November 21, at the end of the ninety-day stay, the county board adopted the ordinance.

When UPA and CPA announced their intention to apply for a permit on November 26, Jerry Kingrey expressed their frustration with the delay: "When REA first brought electricity to rural America, some held out and were satisfied with kerosene lamps. That's why we have eminent domain—to prevent this from happening."

The battle shifted to the planning commission's hearing on the permit application. The six planning commission members were in a difficult position. At these meetings they faced the power companies on one side, armed with legal and technical expertise, and on the other side as many as three hundred angry farmers, intent on stopping the line.

On February 26, 1975, the planning commission voted to advise the county board to approve with restrictions a special-use permit for the CU line. Chairman Sidney Ellingson conceded his vote for the permit would not be popular in the community, but explained that a denial of permit would only lead to an expensive court battle.

Throughout the hearing process, the power companies had made it clear that they would take legal action if denied the permit. John Drawz, attorney for CPA, assured the planning commissioners at the February 26 meeting that the companies "would not roll over and leave the county . . . but rather take a denial to district court." He pointed out that in his experience court reviews were lengthy and costly.

Drawz also emphasized that the companies would oppose any restrictions or modifications that would cause delay. If the permit were denied or if construction were delayed, the co-ops, he said, would have no other choice but to go to the State Public Service Commission and ask for interruptible service for the Grant County area. The farmers charged that threats of legal action and interruptible service amounted to "blackmail." CPA's Jerry Kingrey responded that the only blackmailing done at the meeting was by the farmers, who on one occasion rose to their feet and loudly proclaimed "No" in a straw vote on whether they wanted a powerline to come through Grant County.

In the end the permit was approved, but the planning commission recommended twelve restrictions. Most of them were minor, but one, originally suggested by Jim Nelson, caused the companies concern: "CPA-UPA shall assume liability for damage from ozone, NOx [nitrogen oxides], induced charges and induced currents caused by the transmission line to vegetable, animal and human life." The companies had consistently refused to accept such liability. Otherwise, they had reason to be pleased.

In retrospect, Jim Nelson feels the farmers were never really able to enlist strong support among local officials: "In Grant County, the county commissioners generally felt that the problem wasn't that serious, that lines were built before. They knew people were pretty much concerned about it and they wanted to address those concerns as much as they could. But they basically wished that the problem would go away." He points to the utilities' legal leverage on county officials: "They were more afraid of the power companies' money than the money we could raise. They felt that they could get sued by better, more, higher priced lawyers if they ruled against the power companies than if they ruled against us."

Lee Pederson, who was chairman of the county board, disagrees with Nelson. The legal threat, he contends, was not important: "We're used to having people threaten court action; if the county thought it was the right thing to do to deny the permit, it would have done so." Pederson feels that the important question was whether the line was needed, and county officials were at a real disadvantage in deciding this question: "The farmers need the electricity, but I am not sure the power companies do, and how do you decide? You have to be a damn Philadelphia lawyer to figure out some of the professional language. That's one reason I'm out of public office now. You don't have the information to make these kinds of decisions."

The planning commission was to present its recommendation to the county board on April 7. But by that time it no longer mattered; events in neighboring Pope County had fundamentally changed the course of the protest.

Grant and Pope are both agricultural counties, populations five thousand and eleven thousand respectively. The farms are larger in Grant County, averaging around 400 acres, with more emphasis on crop farming—primarily wheat. Pope County farms average 240 acres, the major crops being oats and corn, with more of a mix of crop and dairy farms. Pope County is blessed with an abundance of attractive lakes. Glenwood, the county seat, is on beautiful Lake Minnewaska, a thriving tourist town during the summers.

What was significant for the CU project was that the Pope County shoreland was zoned, and it was almost impossible to build a powerline across the county without passing through shoreland areas. In this way, Pope County officials had authority over all major utility projects, including the CU line.

In 1971 the Minnesota legislature passed the Shoreland Management Act, which mandated counties to develop shoreland management programs. In January 1972, the Pope County board hired Michael Howe to develop such a program. Howe, a graduate of the University of Minnesota with a degree in wildlife management, was twenty-four when he was appointed county planning and zoning administrator.

One of the first things Howe did after arriving in Glenwood was to write a shoreland management ordinance for the county. In drafting the ordinance, he worked closely with the eleven-member Pope County planning commission, a brand new advisory body just appointed by the county board. One issue was whether to require a special-use permit for utility projects. The utilities wanted to be exempted. Howe remembers regular visits from a representative of Northern States Power, the major investor-owned utility in Minnesota: "This NSP guy came by about every month. He kept trying to sell his package, which asked for complete exemption. They wanted us to adopt their ordinance. I wouldn't recommend what he wanted—I didn't trust him."

Howe made sure the utility companies were not exempt and a special-use permit could be required. The shoreland ordinance, adopted by the Pope County board in April 1972, contained the following provision:

F. *Exemptions:*
1. The following uses, being essential for the operation of any zoning use district are exempt from all the provisions of this

ordinance and are permitted in any district: poles, towers,
telephone booths, wires, cables, conduits, vaults, pipelines,
laterals or any other similar distributing and/or transmission
equipment of a public authority *except, where the proposed
essential service might significantly mar or alter the natural
characteristics of an area, a special use permit may be required.* . . . [italics added]

At Howe's suggestion, the county board amended the 1972
ordinance in April 1974 to read:

. . . . where the proposed essential services might mar or alter
the natural characteristics of an area, *and upon petition of
concerned landowners to the County Board regarding such
actions, a Special Use Permit may be required by the County
Board.* [italics added]

This amendment gave citizens an opportunity to initiate action
requiring utilities to apply for a special-use permit.

In late May 1974 Michael Howe received an environmental
impact report on the cu line from cpa and upa. Then Jerry Kingrey called and asked whether the co-ops would be required to
apply for a Pope County special-use permit. In his dry, under-
stated way, Howe told Kingrey he thought there would be considerable interest in the county and "it was quite likely a permit
would be required and the companies might as well sign an application and have it on file and ready to go." On June 26, cpa and
upa formally applied for the special-use permit.

Farmers from the area known as Bonanza Valley were the first
residents of Pope County to make it clear they did not want a
powerline across their land. One week after the co-ops applied for
the permit the Bonanza Valley Irrigators Association passed a resolution opposing the line. On July 23, the Association's steering
committee unanimously passed another resolution which began:

Whereas the practice of irrigation is increasing rapidly in
Bonanza Valley and there is a limited number of acres of good
irrigable land to be developed; and the most preferred types of
irrigation equipment cannot move over land that is obstruct-
ed by power line poles or towers or other such obstructions;
and local, state, and federal funds in excess of $200,000 have
been invested to develop irrigation in Bonanza Valley.

Therefore be it resolved that the Bonanza Valley Steering Committee objects to and will oppose the construction of the power utility line as planned which will cross prime irrigable land and restrict the use of irrigation equipment and thus reduce the value of such lands.

Chairman Bruce Thorfinnson called the July 25, 1974, meeting of the Pope County planning advisory commission to order at 8:00 P.M. The first item on the agenda was the powerline, and 150 farmers had jammed into a small hearing room in the Pope County courthouse in Glenwood. Michael Howe since has observed: "There was a very big crowd at the July meeting, and it was big crowds from there on. Every hearing on that project was capacity crowd."

Jerry Kingrey presented a slide show explaining the proposed line, its purposes and uses. He described the way in which farmers would be approached regarding easements and payment for the land, and discussed how various problems that might arise would be handled. Then there were many questions. Thorfinnson was concerned about whether the line was needed. He also requested additional information from Kingrey about where the towers would be located and what they would look like. He later observed: "We came up with very good questions at the first meeting—they had to be, because they were never answered."

Thorfinnson, a high school biology teacher in Glenwood and a strong environmentalist, was suspicious of the power companies' environmental report. In describing Pope County, for example, Thorfinnson found that it did not mention walleyes, the major game fish, and listed strawberries as the major export, which was absurd in this community of crop and dairy farms: "The Commonwealth Associates people who prepared the statement never even set foot in Pope County. They flew over the route in a plane. That showed how much they really looked at it, how much they really cared."

John Bohmer, president of the Brooten bank, testified at the first planning commission meeting for the Bonanza Valley farmers. He submitted a resolution, along with a petition signed by many landowners.

Jim Nelson was also there. Michael Howe remembers Nelson as being very effective at the early meetings: "Jim Nelson really at the beginning was the one who was raising all the pertinent questions. And it seemed like with every meeting the questioning from many of the farmers became more and more fervent."

The planning commission decided that there were too many questions about the permit to settle at one hearing. They voted to hold a special meeting to deal with the powerline on September 5.

Then the political maneuvering began. Down at the capitol in St. Paul, state representative Delbert Anderson, a member of the powerful House appropriations subcommittee that oversees the budget of all state agencies, circulated the Bonanza Valley steering committee resolution and a letter from John Bohmer to all his colleagues in the House of Representatives. Anderson's farm, located outside of Starbuck, was on the proposed powerline route. The village of Brooten in the heart of the Bonanza Valley went on record against the line. Mayor Earl Getschel protested that it would interfere with the town's plans for a new airport. The Glacial Ridge Flying Club from Starbuck protested that their town also planned a bigger airport and the powerline would present potentially serious problems.

Perhaps most effective was a letter from Jon Wefald, commissioner of agriculture, to Philp Martin, general manager of United Power Association, which read, in part:

Dear Mr. Martin:

. . . I urge your review of the proposed routing for the new UPA-CPA power transmission line. . . .

Reviewing the environmental report on your proposed transmission line route, I have observed that not only does it traverse some of the best agricultural land in our state, but local concerns have been raised and are mounting in regard to the possible damage to the irrigation-agricultural production potential in our developing Bonanza Valley.

Local authorities estimate that upwards of 3,500 acres of land in this fertile area will be deprived of the full advantage of irrigation if the present transmission line routing is not changed. This involves only 27 miles of the proposed 400-mile route.

Is there an alternative route? Is there alternative technology for underground high voltage transmission lines that would not void the traveling spray boom irrigation method most practical and economical for Minnesota agriculture? Can the plan be adjusted at this point in time? . . .

Irrigation is one of the important new tools of Minnesota agriculture. A year ago the potential outlook of irrigation was minimal. National and world food needs, the new index of grain prices, and widespread midwest drought have made irrigation suddenly economically attractive as well as practical and necessary insurance against the adverse whims of nature.

In conclusion, I do strongly recommend your consideration for an alternative route through land of minimum agricultural potential. . . .

Sincerely, Minnesota Department of Agriculture, Jon Wefald, Commissioner.

The Bonanza Valley farmers knew how to exert leverage. They were active at the local level by the time of the first Pope County planning commission hearing. Commissioner Wefald's letter carried weight at the state level. Moveover, he sat on the Environmental Quality Council, which made decisions that vitally affected utility interests. Delbert Anderson was a powerful, well-placed legislator, and the co-ops had to consider his viewpoint carefully.

Michael Howe was also organizing. He did not have important state contacts, but as planning and zoning administrator he knew the system and he was able to work closely with local farmers. Howe asked Jerry Kingrey for a list of the landowners in Pope County under the line. On August 1, he received the names. The next day Howe sent the list to Mike Rutledge, one of the local farmers who had expressed concern, suggesting that he circulate a petition calling for a special-use permit.

Howe was proving to be a force the power companies would have to reckon with. He had refused to grant utilities an exemption when writing the 1972 zoning ordinance. At his suggestion, the ordinance was amended in 1974 to establish a petition procedure. He told Kingrey that the co-ops would, in all likelihood, have to apply for a special-use permit. Then he made sure it would happen by supplying farmers the petition and the names of likely signatories. On August 7, the Pope County board, upon petition of the farmers, required CPA and UPA to apply for a permit.

The special meeting of the Pope County planning commission

devoted exclusively to the powerline was held on Thursday evening, September 5. Because of the large audience, the meeting had been shifted from the courthouse to the Glenwood High School auditorium.

Jerry Kingrey began with his presentation. He outlined the CU project and introduced a number of experts. Michael Barningham, from Commonwealth Associates, described how the proposed route was established. Arnold Poppens, also of Commonwealth Associates, assured the farmers there would not be any health and safety problems. Charles Johnk of United Power Association explained the special efforts the utilities would make to solve the problems in areas where the powerline crossed irrigation country.

Then Verlyn Marth spoke. The powerline was for the metropolitan areas, he said. Pope County would be "assaulted and desecrated" by this and other powerlines in the same established corridor, all for the sake of urban people. According to Howe: "Verlyn Marth excited quite a lot of adrenaline flowing and quite a number of people." When the meeting was opened up for questions, there was a barrage from the crowd. How was the route established? Was it possible additional lines would later be routed in the same corridor? What about health and safety problems? Was the power from the line needed by rural people? Who would the line really serve?

John Bohmer, representing the Bonanza Valley Irrigators, recommended to the planning commission members that they deny the special-use permit. The hearing was continued to October 17.

Later that month Jerry Kingrey requested a private meeting with planning commission and county board members prior to the October 17 public hearing. The planning commission turned him down, feeling it would be inappropriate to negotiate behind closed doors.

Kingrey did arrange a meeting with the Bonanza Valley Steering Committee (BVSC) on October 15, and they worked out an informal deal. The route would be moved to a less objectionable part of the Bonanza Valley, and the BVSC would no longer oppose the line.

After eliminating their most organized opposition, CPA and UPA decided to take a hard line with the planning commission. The October 17 hearing was a turning point. CPA attorney John Drawz made a long presentation. His position was that they had no choice but to approve the line. Pope County officials, Drawz told them, had no authority to deny a permit. The strategy backfired. Immediately after Drawz finished, commissioner "Swede" Andert made a motion to recommend denial of permit. The motion carried by a 9-1 vote. Michael Howe has reflected on why it happened: "Up to this point, I really thought there would be a compromise. They were really stupid I think, really stupid in their tactics. I think they really should have known you don't just send a slick-looking city lawyer—like he just came out of a hairdresser—to rural people and tell them what to do. You just don't approach them that way."

The planning commission drafted a resolution at the October 17 hearing as its official position on the powerline. This resolution, submitted to the county board, is a striking document, responsive to rural values that had never carried much weight in the way energy companies had traditionally conducted their business:

To: Pope County Board of County Commissioners:

The following resolution was drafted following the Pope County Planning Advisory Commissions' October 17, 1974 recommendation to deny a Special Use Permit for Cooperative Power Association–United Power Association to construct a 400 kv Direct Current transmission line which is proposed to be routed through Pope County, Minnesota.

This Resolution is hereby submitted as the official position statement of the Pope County Planning Advisory Commission.

WHEREAS, Power companies have traditionally used powers of eminent domain over the rights of rural individuals for the purpose of transmitting massive quantities of power over high voltage transmission lines in the most economical manner to population centers mainly urban in nature, and;

WHEREAS, evidence has not been presented indicating that rural areas will be the largest benefactor of additional power being supplied by the proposed C-U line, and;

WHEREAS, Present evidence indicates that the proposed C-U line will be routed to supply mainly urban population centers, and;

WHEREAS, The Environmental Report prepared by Commonwealth Associates, Inc. of Jackson, Michigan has been found to be drastically erroneous in certain factual material, and;

WHEREAS, The power companies, in designating the route of the propsed C-U line, apparently gave very little, or no consideration to the detrimental effects upon agricultural land use areas, and;

WHEREAS, It appears that very little consideration was made of irrigation potential for agricultural lands in Pope County, including areas outside Bonanza Valley; and that very little consideration was given to increasing usage of airplanes as an agricultural management practice, and;

WHEREAS, It is deemed possible to route such a line through areas of less intensive land use, thereby minimizing the impact upon critical agricultural areas, and;

WHEREAS, Evidence has been presented stating that the effects of high voltage power lines upon the health, safety, and welfare of individuals living and/or working near such lines has not been adequately researched by the power companies or by other researchers, and;

WHEREAS, Certain Rural Electric Cooperatives are presently wasting large sums of money in advertising for more use of power when same Rural Electric Cooperatives could be spending this money on research for future power generation by more efficient and less environmentally harmful methods, and;

WHEREAS, The power companies' present policy on electrical rates suffers an inverted rate structure, thereby discouraging conservation of electrical energy and encouraging more use of electricity since larger volume users obtain a cheaper rate structure, and;

WHEREAS, Present power company policies tend to lead toward further desecration of the environment by increased strip mining, additional pollution-producing power plants and more unsightly towers across the countryside, and;

WHEREAS, Pope County is predominantly rural in nature at the present time, and has many areas that would be considered prime agricultural land; and has many areas with scenic, historical, recreational, and wildlife values, and;

WHEREAS, Pope County is presently free of any extremely high voltage transmission lines and any adverse effects caused by such lines, and;

WHEREAS, Present evidence indicates if approval is given for one corridor for high voltage transmission line, that there exists a strong possibility for additional transmission lines along the same or adjacent corridor, and;

WHEREAS, It has not been satisfactorily established before the Planning Commission that there is actually a need for such proposed power line, now:

THEREFORE, Be it resolved that the Pope County Planning Advisory Commission hereby recommends to the Pope County Board of County Commissioners that the Special Use Permit requested by Cooperative Power Association–United Power Association for the proposed transmission line be denied on the basis that it will be detrimental to the health, safety, and general welfare of the present and future residents of Pope County.

This action shocked the power companies. For as long as anyone could recall, local zoning authorities had always acquiesced when it came to powerlines; utilities were granted eminent domain without a second thought and they ran their lines where they wanted. By this action the Pope County planning commission said that the "public interest" involves more than just getting electricity to the consumer. Other values may also be important. It was responsive to such local concerns as the effect on an agricultural environment, possible detrimental health and safety effects, whether the electricity was really needed, and whether other lines would follow this one. When satisfactory answers were not forthcoming, the commission did something quite unprecedented: it recommended denial of the permit.

There was one more step in the process. The planning commission was advisory to the county board. The board scheduled a meeting on November 6 to consider the special-use application.

The power companies organized, working through their local co-ops. At the November 6 meeting the general managers of the three electric cooperatives serving Pope County—Agralite,

Stearns, and Runestone—all strongly urged approval of the line. However, the board was not comfortable with all the controversy and postponed its decision. It set aside another meeting on December 4 to consider the matter further.

The companies then zeroed in on the county board with their big guns. They had help from the highest levels of state government. They first offered to bring in the director of the Minnesota Energy Agency, John McKay. On November 21, Jerry Kingrey wrote to board chairman Mel Heggestad, "We extend to you an opportunity to hold a public meeting at which Mr. McKay would appear. If you wish to accept this invitation to receive state level information on the question of energy use and distribution regarding the cu Project, we ask that you contact the undersigned [Kingrey]. . . ."

They also enlisted the support of the EQC and the attorney general's office. The EQC chairman asked the attorney general if a county had authority to prohibit construction of a high voltage transmission line. On December 3, one day before the Pope County board was to meet, assistant attorney general Thomas Mattson obliged with an "hypothetical" opinion. In reviewing Pope County's shoreland management ordinance, Mattson concluded:

> The overall impact of these provisions of the ordinance seems, therefore, to be that in the case of a proposed HVTL, or any other enumerated essential service, it must be permitted, but in certain cases its construction may be surrounded with reasonable conditions designed to minimize adverse effects of a significant nature. Where the applicant modifies its proposal to meet such conditions, and does not violate them, the permit must be granted. . . .

> You will, of course, understand that we cannot and do not render our opinion on hypothetical fact situations, and that the facts of a specific case would have a signficant impact on any judicial decision. The answers to your questions presented herein deal with our judgment of the legal principles involved.

The attorney general's ruling was widely covered in the media. From the co-ops' point of view, the timing could not have been better. The county board met the next day. Fearing a lawsuit if they denied the permit, the commissioners voted to refer the special-use permit decision back to the planning commission for further hearings. It expressed the hope that the planning commission could work out some "reasonable conditions."

When the planning commission resumed public hearings on January 31, 1975, the companies took a hard-line approach. They were not in a compromising mood. At the meeting, John Drawz reiterated the co-ops' right of eminent domain. Gene Sullivan, manager of Stearns Electric, stressed the local cooperatives' need for additional power. Arnold Poppens of Commonwealth Associates argued that the revised route would have a minimum impact upon agriculture. He assured the planning commissioners there was nothing to the health and safety fears—these issues had been carefully researched and the line was quite safe.

CPA and UPA were not willing to provide the planning commission with much additional information. An exchange of letters in February between Michael Howe and Jerry Kingrey typifies the responsiveness of CPA-UPA to Pope County inquiries. For example, Howe, representing the planning commission, asked, "What is the stage of completion for the CPA—UPA project at this time? What percentage of easements have been purchased *in each affected county* at this time?" [italics added]. Kingrey's February 28 reply to those questions was simply: "a. Under construction, to be completed by August, 1978. b. The easement acquisition activity in Pope County has not yet begun."

The co-ops also tried to pressure Pope County officials directly. One day, co-op representatives came to Bruce Thorfinnson's home. He recalls, "I thought it was pretty cheap. They came in the yard and wanted to talk to me about the line. I'm supposed to be independent. I wouldn't let them in the house and refused to talk to them."

On another occasion, Thorfinnson was called out of his class at Glenwood High School for a long-distance call. It was Peter Vanderpoel, chairman of the state Environmental Quality Council. Thorfinnson was surprised by what he heard: "His line of questioning suggested his viewpoint, which was not in agreement with what was happening in Pope County . . . that we were hicks, stupid, didn't know what we were doing."

According to Thorfinnson, the power companies hired a Glenwood lawyer to investigate him: "They investigated me, did background work on me. I know because the lawyer told me about his meeting with them. That really disturbed me. What kind of people were we dealing with?" Michael Howe also found he was under investigation; he assumed that since the co-ops were not successful with their direct approaches, "They wanted to get some information to do some mud-slinging."

In the meantime the farmers were organizing. The proceedings in Pope County gave them more time and a glimmer of hope. The battle of Pope County, fought out by determined local officials such as Bruce Thorfinnson and Michael Howe, had by now become less a consequence and more a cause of the farmers' protest movement. Farmers from neighboring counties gathered in Pope County to exchange information and support opposition to the line. The hearings were an inspiration to protesting farmers. As long as the commission stood up to the power companies, they had hope. They did their best to make sure the Pope County planning commission held its ground.

In mid-February, farmers in neighboring Stearns County, the largest county in central Minnesota, a major grain producer and the state leader in dairy production, organized to stop the line. One event was especially important in mobilizing the militant opposition in Stearns County. That was the October 15 agreement between CPA-UPA and the Bonanza Valley Steering Committee. As a result of this private agreement, the line was moved out of Bonanza Valley and routed further north in Stearns County, near the town of Elrosa.

"That was their fatal mistake," according to Virgil Fuchs, leader of the Stearns County protest. Fuchs had heard about the line, as had many other farmers, in July 1974. He knew the farmers in Bonanza Valley opposed the line but never believed it would be moved off the sandy soils to the south, which required irrigation to be viable cropland, and onto the rich, fertile land in his area: "In fact, a couple of guys told me, 'We're going to tell them to put the line out at Elrosa because those farmers at Elrosa don't irrigate anyways.' I said to myself, 'They're going to make them put it on the good ground instead of their sand.' So I told these guys there wouldn't be anybody that foolish to change the powerline from this poor land that doesn't pay taxes to the good heavy farmland.

Nobody would ever do that. I wouldn't even consider it. I laughed at him. It was just kind of a tease—that is what I thought it was. I just said they wouldn't be that stupid to go in the heavy land—the land that supported the community all these years, paid the taxes on the schools, and does almost all the business in the towns. They wouldn't go and put it on that land."

Evidence of the new route came in February 1975. Bruce Paulson, a farm neighbor of Fuchs, attended a township board meeting where CPA and UPA explained their project. They brought a map of the change in route which showed the line going near Elrosa. It was at this meeting that Paulson also realized tht the Stearns County board had already granted approval for the line. He made a copy of the map and immediately called Fuchs.

Fuchs remembers this day, February 15, well. It was the first time he heard the route had really been changed to run across his farm near Elrosa: "I called [county commissioner] Henry Dickhouse in Melrose and asked him what about this powerline that is supposed to be going through Stearns County. Are the commissioners going to allow this? And he said, 'Oh, don't worry about this powerline—it is only going through a corner of Stearns County—it is not going to come close to here.' I thought, 'Here,' and I asked 'Where is here?' He said 'Melrose is here,' and then he asked, 'Who is this calling?' and I says, 'This happens to be somebody where the powerline is supposed to go over the top of.' He says, 'Oh, I am sorry. I didn't know where you are from.' So what that tells me is, 'It is not going to be close to here,' means it is not going to be close to home to him—in thinking, he was representing the people from Melrose instead of representing us out here."

Fuchs and Paulson contacted all the farmers along the route in western Stearns County. Virgil Fuchs recalls: "We got a copy of the route and reroute and I guess that was our strongest point—was to show people what the other people did in choosing us second-class citizens. I had papers along to show the farmers what happened in the past and why they wanted to put the line through here. They called our land here 'land of less agricultural potential' to move it from Bonanza Valley. That's what they called our land. It didn't take too much trouble to stir people up once they'd seen that quote. They'd say, 'Well, they are crazy; who in heck would do that?' "

The farmers asked Fuchs and Paulson what they should do. They were told not to sign any easement agreements and to spread the word to their neighbors. Fuchs felt most of the farmers were responsive: "It worked pretty good. We had some problems though. We were about a day or two ahead of the power companies. A lot of people when we got to them said, 'Oh, no, no, no—there is no line coming through here.' And then at one place when I got home the guy was on the phone—'Hey, hey what am I going to do. They were here this afternoon.' I told him, 'Tell 'em to go to hell.' "

This person-to-person organizing paid off. None of the farmers in western Stearns County signed easements. By the end of February, the farmers were ready for a formal organization. On February 27, 1975, Jim Nelson and Verlyn Marth came down from Grant County for a meeting. Nelson spoke at the beginning. He listed the "bad effects" of the line and concluded, "What we really need to beat this thing is to all work together. The power companies' favorite tactic is to divide the people, and we can't let them do that."

Verlyn Marth warned of "the power cartel people." Rural people, he said, were "just a colony to be used." He outlined what he felt would be the power companies' tactics. They would "try to ram it through—you can't do anything about it, don't try it, we're bigger than you are," "divide you up," and "threaten you with lawsuits." He urged the farmers to get a lawyer and fight the companies: "You're being programmed to think you are helpless. But they are an evil cartel assaulting individual farmers and they can't stand the kind of publicity that says what they are. It is your responsibility to beat the line. You are the stewards of the land."

The farmers at that meeting voted to organize Keep Towers Out (KTO) and fight the powerline. Mary Jo Paulson: "Your heart would just bleed that you knew this was happening and you just had to get people to see it." Their first task was to find out about the line from "people in the know," to prepare for the utilities' first public information meeting in Stearns County, March 14 in Paynesville. Jim Nelson was especially helpful. He provided Stearns County farmers with reading material and the names of other people to contact. Included in the reading was Louise B. Young's book, *Power Over People*, Nelson's bible for powerline protest. Virgil Fuchs also did a lot of digging on his own. Mary Jo Paulson said: "If Virgil thought of someone who might know facts about the line, he called them up right away. Virgil's phone bill, oh, man!"

The farmers in Stearns County spent a great deal of their time supporting the farmers in Pope County, where there was, they thought, a real chance the line could be stopped. There, they linked up with another new organization, Towers Out of Pope Association (TOOPA). As CPA and UPA pushed harder for local approval in January and February, farmers in Pope County felt the need to coordinate strategy above and beyond individual testimony at public hearings. In early February, 150 farmers, most living along the proposed route, formed TOOPA. Harold Hagen, one of the most outspoken farmers at the public hearings and an opponent of the power companies, was elected chairman.

Hagen was an articulate, politically sensitive spokesperson for the farmers. He emphasized that not only did the powerline stand as an obstacle to farm work, it threatened to take valuable land out of production or decrease productivity. Such results would affect many more people than just Pope County farmers. What happened to farmers was of great consequence because of the special commodity they produced—food. A threat to their ability to produce was a threat to everyone who lived in the country. The rhetoric—sincere rhetoric—of saving the good land for food production became a familiar theme of the powerline protest.

There were also other concerns that brought farmers together and propelled the rapidly growing protest movement. Virgil Fuchs expressed the special feeling many farmers have for their land: "We've spent almost all our years farming—in developing the land—like this farm here was all holes and pastures, swamps —it is probably twice as good as when we started because we drained out the holes and cleaned up the place, planted new trees, took fences out of the fields that were old and shabby, and drained out the land. Those first years when we were farming here—you know when you get married you have a lot of friends that get married—heck, we didn't go to half of the weddings because you stay home and you work and work and work day and night to keep going. You put everything you have into the place. And then to have somebody come along after you do all that—that was I guess why I started. [It] was the idea that a guy can just stay at home and work and work and work and all of a sudden somebody can just come along and take it away from you—I guess that was my thinking. Before that I kept to myself, left everybody else alone and they left me alone, more or less. I didn't get involved in anything that was going on."

For Jim Nelson, "There would be three things: the health and safety, the basic problem of farming, and then the question of whether this is something we really should be doing to the countryside. I think that's a pretty important one because farms have generally been in a family for, in some cases, several generations and, in some cases, since the land was homesteaded, and they feel like they've taken pretty good care of the land, and they just don't feel that this is something that should be done to it. I think one of the reasons health and safety is a major concern is that technically, say if it ruined a quarter of my land or something like that, you can be compensated for that. But you can't be compensated for having an electric field or ozone or something cut ten years off your life—you can't be compensated for that. I think that's why it has been a major concern, because it's something that just by its very nature you can't be compensated for. The other thing you can't be compensated for is this basic feeling of being a farmer. The farmer has a sacred trust in protecting that land and preserving it for future generations and that's another feeling along the same line as health and safety that can't be compensated for. If you've let the land, because of the powerline, decay or basically be destroyed to some extent, it's not something that a person can be compensated for financially. So there are three basic questions —the farming, the health and safety, the sacred trust idea—and two out of the three are out of the realm, logically, for financial compensation."

Verlyn Marth believes that "the powerline was a massive attack on the one thing that farmers have any real strong commitment to and that is their land. The old land use—a fundamental issue in human relations is ownership of land. I'm not talking about the economic one your lawyer friends talk about, I'm talking about the land morality, the thing that they are the stewards of the land, and that they have assumed the protection of the land, and that they have done it for years or generations or months depending on the farmer, and they are the stewards, and outsiders are desecrating it with this massive thing against their will. And only that would have caused the protest. Now that was not sensed by management, the depth of feeling—that's the fundamental thing."

The farmers in western Stearns County were outraged that the line had been moved onto their land, and the public information meeting held in Paynesville the evening of March 14, 1975, did little to lessen their anger toward the power companies.

The farmers asked pointed questions about the line. They wanted to know why the power companies could not assume full liability for health and safety problems. UPA's public relations director, Don Jacobson, responded, "That's a legal question and I don't feel qualified to answer that." Why not an urban area? they asked. CPA's Jerry Kingrey answered, "We don't like to route a transmission line through a highly populous area because it affects more people . . . the idea of routing a transmission line is to affect the least number of people." It was a mistake, and before he could change it one of the farmers in the audience yelled out, "Farmers don't mean nothing!" There was a long round of applause.

After that meeting an incident occurred that the farmers still talk about. Ben Herickoff, aged seventy-one, walked up to Don Jacobson. He recalls what happened: "That evening, that Jacobson, he sure got it. I told him at that meeting, I tell you this much —the first one that puts a foot on my land, he is gonna get shot, and I hope that you are the first one."

Later the power companies rerouted the line a couple of miles to the north. In the folklore of this protest, that move was attributed to Ben Herickoff and his encounter with Don Jacobson.

On March 21, farmers from Grant, Stearns, and Pope counties met in Glenwood to establish a multicounty organization. At Jim Nelson's suggestion, they called it CURE—Counties United for a Rural Environment. They hoped a united effort in Pope County would stop the line.

Representatives from each county sat on CURE's board of directors. Jim Nelson was asked to serve as president, but he declined, saying that he needed to spend more time on the farm. Harold Hagen, from Pope County, was elected instead.

The farmers wanted to "impress" the power companies. Their first step was to hire a lawyer. Bruce Paulson from Stearns County had heard of one lawyer, Norton Hatlie, from a friend who managed a nursing home. Hatlie, Paulson reported, had some experience in fighting utility companies.

The founding meeting of CURE concluded with a discussion of leverage and power. During that discussion, Chuck Mexel, a Pope County farmer, made a statement that proved remarkably prescient, in light of subsequent events. He warned his neighbors not to count on "the facts" or the courts to win their battle: "Let's fight a legal battle, that's fine. But let's also fight a political battle —that's where it will be decided, in mass meetings and public indignation. If people do not become angry, all the legal battles in the world won't be of much help. I am sure of it. I'm not saying we shouldn't pursue the legal process, but I know lawyers and I know judges and I know the judicial process in this country and it ain't good."

If the farmers hoped to stop the line by supporting the efforts of the Pope County planning commission, however, those hopes were suddenly, unexpectedly, totally dashed. On March 21, 1975, the day CURE was formed, Michael Howe sent a letter to Jerry Kingrey advising him that the planning commission was considering three possible actions regarding the special-use permit— approval, denial, and tabling: "Regarding possible approval of this permit, the Commission is of course considering stipulations or conditions of the permit that it 'considers necessary to protect the public health, safety, and welfare or necessary for compatibility with surrounding uses and environs.' " Howe outlined fourteen conditions and asked CPA and UPA to indicate whether they would comply with them.

Two weeks later, at a dramatic meeting of the planning commission in Glenwood on April 2, Kingrey rose to reply. One by one he read the conditions aloud and announced the companies' response:

> Condition #1: Preliminary approval until exact alignment of HVTL can be indicated with towers spotted at locations of least inconvenience to landowners.
> Response #1: "Condition Number One is not acceptable

to the cooperatives. Preliminary approval is not acceptable if there is any chance of a later disapproval. I think that the reason for that is obvious—we can't, we couldn't, afford to spend the time and the engineering for a later denial."

Condition #2: Condition that landowners be given an initial easement payment and an annual rental fee based on current market value—subject to rate re-negotiation at periodic intervals of five (5) years.

Response #2: "Our response to condition Number Two is that CPA-UPA cannot spend the dollars necessary to complete this project with a limitation of their rights to continued operation."

Condition #3: Condition that CPA-UPA submit a detailed plan stating how the construction is to take place, and how the affected lands are to be restored to their original state, and how the affected landowners are to be compensated for damages by construction and maintenance crews.

Response #3: "CPA-UPA response is, 'That is acceptable.' "

Condition #4: Condition that CPA-UPA submit complete engineering plans and long-range operational plans stating exactly how the HVTL is to be used (i.e., maximum voltage to be carried by the conductor, etc.).

Response #4: CPA-UPA finds that acceptable."

Condition #5: Condition that CPA-UPA include as part of the easement instrument a statement stating that no additional easement be given, or no additional structure or structures be allowed on the same right-of-way until 75% of the affected landowners sign a statement of no objection.

Response #5: "CPA-UPA finds the condition Number Five —that they cannot be committed to inaction of any sort based on the approval of a percentage of landowners. The power plant siting act henceforth supercedes all local regulations on high-voltage transmission lines."

Condition #6: Condition that before construction begins, CPA-UPA be required to furnish a written statement from at least 75% of the landowners affected stating no objection to the line.

Response #6: "CPA-UPA finds that unacceptable. The completion of the project cannot be subject to the whims or caprice of a percentage of landowners in any given area."

Condition #7: Condition that damages will be paid to individuals affected by the HVTL even though their real property has no actual easement for the HVTL.

Response #7: "CPA-UPA finds Number Seven, ah . . . not an automatic payment—we could not be committed to an automatic payment under those circumstances. If the landowners . . . landowner proves diminution in market value, he has a right under Minnesota law to claim recovery or damages."

Condition #8: Condition that landowners be given an option to choose between lattice steel towers and the single-pole type structure.

Response #8: "CPA-UPA finds condition Number Eight unacceptable—subject to caprice."

Condition #9: Condition that a larger conductor diameter of 1 and ¾" be used (as recommended by Dr. Merle Hirsh of University of Minnesota at Morris) to minimize the corona effects of the proposed line.

Response #9: "Condition Number Nine is found by CPA-UPA to be unacceptable—our engineering research has long since shown our present design to be the best from all interests."

Condition #10: Condition that CPA-UPA offer to compensate for the partial "taking" of landowners' rights by purchasing (at a fair market price) or to relocate (to a desirable distance) all residences and buildings within a distance of five hundred feet of the centerline of the HVTL. This offer "to purchase or relocate" should be extended by CPA-UPA prior to the construction of the proposed line and should remain in effect for at least one year after the line is put into operation and energized to +/−400 kv. Said obligation to purchase or relocate buildings shall be at the option of the property owner and same property owner may waive these restrictions completely if he so desires.

Response #10: "CPA-UPA finds that unacceptable, arbitrary, and it does not relate to any goal of the shoreland management ordinance."

Condition #11: Condition that CPA-UPA accept absolute liability for damages caused by the HVTL, whether damage be due to radio interference, television interference, induced currents, audible noise, the production of ozone, or biological effects of electrical and magnetic fields.

Response #11: Response Number Eleven . . . unacceptable as it relates to absolute liability. Again, state standards establish limits."

Condition #12: Condition that CPA-UPA make up the difference for extra charges made (if any) by aerial applicators to farmers for any inconvenience caused by high-voltage transmission lines.

Response #12: "Response Number Twelve . . . unacceptable, subject to caprice. 'Extra' charges based on what? . . . would be a complete open-end deal."

Condition #13: Condition that CPA-UPA submit a detailed evaluation based on the Minnesota Environmental Quality Council criteria for locating HVTL, showing that the present route is the best route as compared to the twenty-seven (27) possible routes which the *Environmental Report* mentions were identified; and compared to other possible routes mentioned in previous discussion with the commission.

Response #13: "Response Number Thirteen—this evaluation of our route is contained in the Federal Environmental Impact Statement, and supplementary data thereto."

Condition #14: Condition that this CPA-UPA HVTL may not exceed a voltage of 400,000 volts and that CPA-UPA install a monitoring device in Pope County which indicates exactly what voltage is being carried on said line at any given time.

Response #14. "Response Number Fourteen . . . as we have noted numerous times, our DC line will be +/−400 kv— carrying a positive 400 kv pole and a negative 400 kv pole. The monitoring device in Pope County is not a reasonable condition."

The companies thus rejected outright eleven of the fourteen conditions. It seemed a confrontation was inevitable. Very likely the planning commission would recommend the conditions; if the board adopted them and the companies refused to honor them, the line would be stopped. The farmers had won!

But then Kingrey sent a shockwave through the crowd. He announced:

Cooperative Power Association and United Power Association have submitted an application to the Minnesota Environmental Quality Council for corridor designation under the Power Plant Siting Act. This procedure and the subsequent procedure of routing within the designated corridor will take an established period of time. Because any further delay could mean hardship in the future, because of insufficient electrical power for more than one million people in the state due to increasing costs of construction, it was decided additional delay could not be tolerated if the project was to meet essential deadlines. This led the cooperatives to conclude the decision-making process must be shifted from the county level to the state. All local zoning, building, and land use rules and regulations are superseded and preempted by state action. We then leave tonight's decision to you. The formal recommendation of approval of the best and alternative routes will be considered by the power plant siting staff in reaching their conclusion. If you feel a route is the best route across Pope County, you should recommend it while you have the authority. Thank you.

Thus, according to Kingrey, the planning commission had two choices—accept the line on the companies' terms or lose jurisdiction.

The commissioners were stunned. According to Chairman Thorfinnson, "We thought they were stuck with us." The Pope County planning commission voted unanimously to recommend denial of the special-use permit. But it was clear the action no longer had meaning.

They claim that the people have a right to power
But only the electric kind
We're supposed to be quiet while they build their towers
'Cross the land that's yours and mine

—from "Minnesota Line" by Nancy Abrams

3   The Public Participates:
The State Decides

The Question of "Where?"

On April 1, 1975, the day before his dramatic confrontation with
the Pope County planning commission, Jerry Kingrey had already
set into motion a process that would remove siting authority
from the county and give it to the state of Minnesota. On behalf of
CPA and UPA, he had submitted an application to the Minnesota
Environmental Quality Council (EQC) requesting it to assume
responsibility for siting the powerline, under provisions of the
Power Plant Siting Act of 1973, from which the co-ops had previously chosen to be exempted.

Appended to the application was Engineering Report R–1670,
prepared by the co-ops' principal consulting firm on this project,
Commonwealth Associates of Jackson, Michigan. In his covering
letter to the EQC, Kingrey noted than an environmental impact
statement (EIS) would be required, requested that it be coordinated with the siting process, and recommended "that the EQC
select its Power Plant Siting Staff as the Responsible Agency for
the preparation of the EIS."

On the very same day, April 1, 1975, a young man came to work
for the EQC. His name was Larry Hartman. He had earned a master's degree in urban and regional planning from the University of
Michigan in 1972 and he knew quite a lot about high voltage
transmission lines. His last major assignment was a five-hun-
dred-kilovolt line in Canada. His employer had been Commonwealth Associates. He now joined the Power Plant Siting Staff of
the state of Minnesota.

On April 8, the EQC accepted the CPA-UPA application. Assigned
to the project were Larry Hartman and another newcomer to the
Power Plant Siting Staff, George Durfee.

The CU powerline thus became the first transmission line to be
considered by the state of Minnesota under provisions of the new

laws that the state legislature had enacted in the wake of the environmental movement and the Arab oil embargo of 1973. As already noted, these laws had created two new agencies, the Energy Agency and the EQC.

Under provisions of the Minnesota Energy Act of 1974, the Energy Agency director had to issue a certificate of need, certifying that a proposed new facility was "needed." Only after a certificate of need had been granted could a utility apply to the EQC for siting of the facility. The Energy Act was passed on April 15, 1974, but when CPA and UPA approached the state a year later, the Energy Agency was not fully prepared; the rules and regulations governing the certificate of need process had not been completed. After consulting with the attorney general's office, the EQC decided to accept the application and begin the siting process even though a certificate of need had not been issued. It did stipulate that the need determination would take place later and therefore the application would be accepted "at the risk of the applicant with regard to the processing of a Certificate of Need by the Minnesota Energy Agency." Thus the state would decide *where* the powerline would be built before it decided *whether* it should be built.

The EQC then ventured forth into unknown territory with its first attempt to route a powerline. It was guided in this effort by procedures and values set forth by the legislature. The process to determine a route would have two six-month parts. First a *corridor* up to thirty kilometers wide (about twenty miles) would be selected. Then the actual *route* the line would follow would be chosen within the corridor.

The prescribed procedures for both corridor and route selection stressed *public participation.* Peter Vanderpoel, then chairman of the EQC, described the new mechanisms for citizen input:

> . . . there are tremendous opportunities for public involvement, public influence . . . [T]he EQC does three things:
>
> 1. It names a citizens' committee to study the proposals and choose which is the least undesirable. . . .
>
> 2. It holds public *information hearings* in the areas that will be affected to explain the power company proposal and alternatives.

3. It holds a series of *public hearings* in every affected county. They are conducted by an *independent* state hearing officer who is *not* connected to the EQC. At these hearings, any citizen—any affected farmer, any citizen, any citizens' group such as CURE—is totally free to offer facts and arguments and recommendations. Then they are on the record, and they must be considered by both the hearing officer and the EQC.

On paper this process offered an unprecedented opportunity for citizen participation, a real departure from past practices. In Vanderpoel's words:

In the old days, the companies decided where they wanted to go. The companies made those decisions by themselves—in their corporate boardrooms. No citizens—no farmers, no small businessmen, no workers, no county commissioners— were involved in those decisions. The companies had the power of eminent domain—a power the Legislature has not given to most state agencies. The power companies could condemn land for their towers. *They* decided where they wanted to go. Then they negotiated quietly with each individual landowner—and there was always the threat of condemning their land, if the farmer didn't sign on the dotted line.

In 1973, the Legislature changed that. It decided the government should have some control over the power companies. It established the EQC. And one of the jobs it assigned to the EQC was power plant and line siting. . . .

The Legislature also said the public should have a hand in this process. So now, the EQC has a series of procedures that guarantee public involvement—more, I believe, than any other state agency. Now citizens *do* have a voice, and they *can* make a difference in where powerlines and plants go.

Narrowing the Scope of Inquiry

Before the public was allowed in, however, some crucial decisions had already been made. Which criteria would count and which would not count in siting the line had been addressed by the legis-

lature; the scope of possible siting choices had been limited by the
EQC.

The siting criteria mandated by the legislature clearly reflected
the impact of the environmental movement. The Power Plant Siting Act of 1973 categorically excluded national and state wilderness areas from high voltage powerline routes. It also specified
"avoidance areas," including parks, wildlife areas, wild, scenic,
and recreational rivers, and designated trails, canoe routes, and
boating routes. Throughout the public hearings on the powerline,
many of the concerned farmers, acutely sensitive to the importance of agricultural land, food production, and rural people,
chafed at the values implicit in this listing.

Before the hearings began, the EQC limited the questions that
would be considered. As shown in figure 3, the corridor application from CPA and UPA designated one preferred corridor (I) and
one alternate corridor (II). Both corridors entered Minnesota from
the west at the same point on the North Dakota border and terminated at the same point, near Delano, Minnesota, just outside the
Twin Cities. As far as CPA and UPA were concerned, these end
points were fixed. While seeking approval from Pope County
officials, they had proceeded to acquire easement rights by negotiating quietly with individual landowners along their preferred
route in North Dakota and in Traverse County, Minnesota,
which borders on North Dakota. They naturally preferred to enter
Minnesota along those easements. They had also acquired land
near Delano, where they planned to build a substation to convert
the DC power to AC for distribution in Minnesota. The CPA-UPA
application for corridor designation stated, "The study area
should be viewed having reference to the Minnesota entry point
and the location of the conversion facility near Delano, Minnesota. The Minnesota entry point is fixed by the exit of the line
route from North Dakota. The conversion facility site has already
been acquired and site development commenced."

At the June 10 EQC meeting, the staff recommended to the
council that two additional corridors (III and IV) be added and that
the study area for the line be limited to the four corridors designated. As shown in figure 3, the new corridors each began and
terminated at the power companies' preferred end points; III filled
in the area between corridors I and II; IV extended the study area

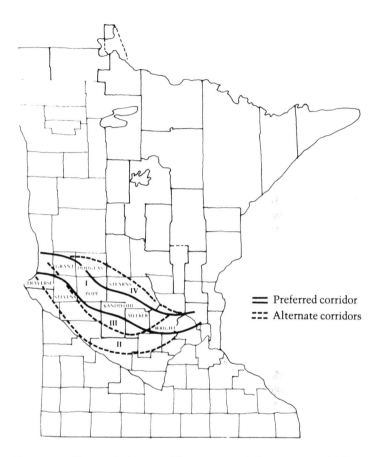

Figure 3. Transmission corridors proposed for CPA-UPA high voltage transmission line

to the north. This was significant in that it brought interstate highway 94 into the study area. In adopting the staff recommendations, the EQC effectively limited the scope of subsequent inquiry. The corridor selected would have to be within the study area. The end points were fixed.

At the same meeting, the EQC adopted another staff recommendation, which further limited the routing possibilities: it excluded all municipalities from the DC powerline route.

At this point, after the scope of the inquiry had been narrowly defined, the public was invited to participate. EQC chairman Vanderpoel has presented his perception of the public process: "In June of 1975 we held information meetings about the corridors in nine counties—Renville, Swift, Pope, Wright, Stearns, Todd, Grant, Hennepin, and McLeod. That's four more than the law requires. They totalled about twenty-eight hours.

"Next, we held public hearings in all those counties in July. They totalled seventy hours. Citizens testified until they were out of words.

"And a 47-member citizens' committee devoted more than 651 man-hours to meeting and recommending a corridor to the EQC.

"We accepted recommendations of the hearing officer and the citizens' committee and picked one corridor for more study.

"Then we held information meetings in each and every one of the twelve counties in the corridor, totalling about thirty-two hours. Next were public hearings totalling about sixty-one hours.

"The 48-man citizens' committee met thirty-one times totalling about seventeen hundred hours of public involvement.

"Does all this public involvement make any difference? I think so—I think it has. . . ."

That is one perception of what happened. The farmers have another. It is revealing to look at "public participation" from the perspective of the "public" who "participated."

To begin with, the farmers who opposed the powerline were deeply upset, but not altogether sure what to make of the state's intervention. They were angry because to them it seemed the rules of the game had suddenly been changed just when, playing by the rules, they had won in Pope County.

Verlyn Marth recalls, "We were really suspicious. . . . They came in and told us that we were not under the EQC, they were grandfathered out. Now, the law being unchanged, they come back and they say now we are grandfathered in." Jim Nelson remembers having mixed feelings at the time. On the one hand he realized "The power companies saw them [the state agencies] as kind of a bailout. They were in trouble and the state agencies would get them out of trouble." Yet at the same time, he recalls, "When we first heard about it, we thought we'd won a little bit because we'd made them go to a state agency and they were in trouble with the county commissioners, and we assumed that since they didn't go to the state agency in the first place, it must be worse for them there than with the county."

It was with this mixture of anger and hope that the farmers geared up to continue their fight in a new forum.

The Hearings Begin

The EQC process began on June 16, 1975, with two weeks of evening "information meetings" in towns scattered throughout the study area. The official purpose of these meetings was to provide an opportunity for the EQC staff and utility representatives to meet informally with the public, explain the project, answer any questions, and generally make sure that people were informed before the formal public hearings that would follow. The protesting farmers worked to get local people to the meetings and many cut short their farm work to follow the meetings from town to town. Virgil Fuchs was at all of them with his tape recorder. He recalled in an interview how they were organized:

Fuchs: "Jerry Kingrey and Don Jacobson, the power companies' PR people, were at the information meetings and you didn't have to be sworn in to say something and there was no court reporter keeping records. I got records of all the information.

Q: "But those meetings were under the auspices of the EQC and Larry Hartman presided at them?"

Fuchs: "Yes. He was there and you'd ask a question and then he would direct it to "Well, Jerry, would you like to respond to that?" and then he just kept order. . . . They weren't supposed to be running it, but if you were at one you could sense it. Larry Hartman would look at the power company when something would come up and then the power company would answer it. The power company ran it more than the EQC. They had all the answers; the EQC didn't know anything."

Fuchs got the impression that the information meetings were being used less as a device for informing the public and more as a scouting expedition in preparation for the public hearings. In his words, they were like "finding out how deep the water was and then come back next week with your hip boots and your spear and go fishing," or "going out in the woods and seeing how many deer there are up there and the next week the deer hunting season opens."

But if they expected fish or deer, they were disappointed; what they found instead were wolves.

Fuchs: "At every one of these information meetings, the people would tell them . . . 'You come across our property, you're not gonna be very welcome.' One of the real things that people would be concerned about was liability, and the farmers would ask them who was liable if a tractor runs into and damages a tower? Who is liable? And the power company would say, 'Well, in the past the landowner has been responsible.'

"We knew at the time there would be some shock . . . they would say there would be no shock . . . we'd try to establish that there would be a little shock and ask where do you draw the line? You know, if it kills, or it don't kill you? If it knocks you on your butt? . . . these kinds of things, you know, and get the whole crowd kind of thinking. . . .

"What happened at every one of these information meetings was, I guess by the time it was over with, we had the whole crowd so against them that they just about got booted out of the place, almost every one of them. Everybody would get so worked up it would be at the point of a riot when it would be over with."

Q: "How many people would talk?"

Fuchs: "Everybody'd be hollering."

Next came the public hearings. In contrast to the information meetings, the corridor hearings were more like a court proceeding, complete with a judge (the hearing officer) and sworn testimony. The hearings opened in the Renville County courthouse in Olivia, Minnesota, on the evening of July 21, and continued almost nightly for two weeks, in the same towns where the information meetings had been held.

The hearings were presided over by Harold Evarts, a Minneapolis attorney. His job as hearing officer was to conduct the hearings and prepare findings of fact and recommendations for the EQC. On that hot summer evening in Olivia, he promised that the hearings would continue "until all interested persons, representatives, and organizations have had the opportunity to be heard by submitting oral or written data, statements, or comments, or by presenting witnesses."

First to testify were the power companies. The thrust of their testimony was technical. John Drawz, CPA's lawyer, began by submitting the corridor application and several technical documents prepared by Commonwealth Associates for the record. He then introduced testimony by a series of expert witnesses from Commonwealth Associates about such technical topics as electrical and environmental impacts of the high voltage transmission line, the prospects for running such lines underground, and the scientific, objective manner by which the preferred route was chosen.

To Commonwealth Associates, routing the line was a numerical problem, with quantifiable constraints (see figure 4). The documents submitted describe how the route was determined:

> The search for an environmentally compatible transmission route began with the selection of a suitable study area by a multidisciplinary team of planners, biologists, landscape architects, foresters, and ecologists. . . .
>
> Information regarding the natural and cultural environment of the study area was collected and systematically organized to provide a general overview of the region. These data concerning physiography, water resources, climate, vegetation, soils, wildlife and land use were analyzed by the specialists of the project study team. . . .
>
> From the original study area a primary study area consisting of 36,240 square miles in North Dakota and Minnesota were delineated. . . .
>
> The primary study area was then delineated on 1:250,000 scale USGS topographic maps and a detailed inventory was made. Environmental and cultural variables were identified and rated according to their relative importance as constraints to transmission line location.

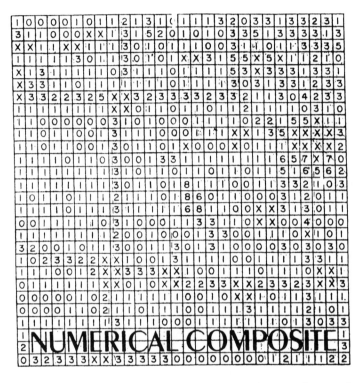

Figure 4. Numerical technique used to route the powerline. Each square mile area on the map was assigned an avoidance rating; the higher the number the more it was to be avoided in routing the line. For example, state lands rated 5, interstate highways 4, and forest land 3. Airfields, state parks, federal land, and lakes were excluded entirely (x on the map). Farm land was rated 0.

A grid system of 36,240 one square-mile grid cells was used to facilitate the inventory. Each grid cell was given a composite rating reflecting both the frequency of occurrence of the constraint variable present and its corresponding impact rating. [See figure]

The constraint variables inventoried included:

Transportation: Airfields, Interstate Highways, U.S. Highways, State Highways

Water Resources: Lakes, Rivers, Marshes, Bogs

Ownership: Federal Lands, State Lands

Vegetation: Wooded Areas

Land Use: Incorporated Areas

. . .

A search was made within each area of compatibility for suitable alternate line routes. These alternates were then transferred to county highway maps.

Each of the alternate routes was evaluated by the professional staff for suitability. Future land use, soils, zoning, and land ownership along each alternate were examined using county plans, zoning ordinances, soil surveys, and plat books. This was followed by an intensive field inspection by the members of the study team.

The alternates were also evaluated and adjusted in terms of visual and aesthetic impact during the field inspection. Upon completion of the field inventory, an aerial inspection of the entire primary line was conducted.

Commonwealth experts testified at length as to the wisdom of the proposed line. Jess T. Hancock, Jr., electrical engineer, assured the hearing officer that ozone, nitrogen oxides, and audible noise would be no problem, electrical shocks could be eliminated with proper grounding, television and radio interference could be corrected, and concluded, "the effects of the proposed +/−400 kv DC HVTL should not be inconsistent with the safety or welfare of man." Carl Haines stated that undergrounding would cost seventeen times more than overhead wires and hence was prohibitively expensive. Michael Barningham, environmental coordinator of CU line, explained how the preferred corridor was selected and why it was best. Arnold Poppens, project manager, emphasized how far along the project was, noting that construction of the powerline would commence "next spring, possibly even this fall yet," and cautioned that failure to choose corridor I or corridor II would result in significant delay.

Cross-examination of these witnesses was abortive at best. The farmers simply didn't have the technical personnel or resources to perform comparable detailed analyses and challenge the assertions of the "experts." The EQC staff treated these witnesses with kid gloves, rarely cross-examining at all.

All interested state agencies were invited to present testimony at these hearings. Naturally they sought to protect their own turf. For example, the highway department testified at the Olivia hearing. Citing criteria laid down by the American Association of State Highway Officials, the commissioner of highways warned against a route paralleling interstate highway 94:

> The visual quality in view of highway road sites and adjacent areas can be materially affected by the size and type of utility facilities permitted occupancy. Consequently, above-ground installations are not accommodated in scenic strips, overlooks, rest areas, recreation areas and rights-of-way of highways that pass through public parks or historic sites. . . . I-94 is designated as a scenic area from Osakis to Fergus Falls [over fifty miles] because the view from the freeway passes two rolling parks, ponds, and lakes near streams and rivers, which is appealing to the eye of the traveler. . . . In summation, we feel that the installation of the 400 kv DC transmission line immediately adjacent to I-94 or other high-type freeways would have a negative effect on the traveling public, because, number one, a maximum visual impact to a maximum number of people; two, the intrusion on the scenic highway designation; and, three, legal problems with occupancy on scenic easements. The Department would, therefore, recommend preserving an open area one-quarter mile wide on each side of the freeway and avoidance of listed sites as much as possible.

Other agencies testified, too. For example, the department of aeronautics cautioned that the powerline should not come within three miles of an airport with a runway longer than thirty-two hundred feet or come nearer than a mile of any lake over one-half mile long, where sea planes might land. So highways, airports, and lakes, along with towns, wilderness areas, parks, wildlife areas, and woodlots, were out. Not much was left to consider—except farmland.

At the second public hearing, the EQC staffers assigned to this project, Larry Hartman and George Durfee, took the stand. Hartman presented a series of technical maps, showing detailed features of the study area. Durfee announced that the staff "will not present an evaluation of the relative merits of the four corridors at this time." However, he continued, the staff would reserve the right possibly to make a recommendation to the hearing officer near the close of the hearing process.

Some of the farmers wanted to know if they could propose further corridors beyond the study area. The answer was no. EQC staffer Durfee, Hearing Office Evarts, and assistant state attorney Jean Heilman made that clear:

> MR. DURFEE: It is my understanding, from a resolution adopted by the council earlier, that they could not. The council said that a certificate of compatibility may be issued for corridors I through IV as indicated or a combination of these corridors. That would seem to preclude consideration of additional corridors. . . .
>
> MR. EVARTS: . . . I am limited in the scope of this hearing to the direction given by the council. . . .
>
> MS. HEILMAN: . . . The council has designated the study area, has designated the two fixed points, the entry point and the termination point. That is what the public is expecting in response, not to enlarge the corridors or to change the entry point. So, I think questions in that regard are not going to be of value in this hearing.

The Farmers' Perspective

By this time, many of the farmers were becoming uneasy about the hearings. For one thing, they *looked* set-up. The hearing officer, Evarts, was up in front of the room, flanked on one side by the EQC staff, Hartman, Durfee, and Heilman, and on the other by the power company lawyers, Drawz and Miller. They would confer, chat, and joke with each other, while the farmers would be in the audience feeling like outsiders looking in.

Bruce and Mary Jo Paulson recall what made them feel uneasy:

Bruce: "Well, first, Larry Hartman; we found out he was working for Commonwealth Associates when they did it. So that
looked suspicious. Secondly, one of their assistant attorneys,
which should have been a representative of the people, she was
hired by the state, paid by the state, and the utilities had two of
their lawyers there, and she should have been representing the
people and yet she was right in with them like flies in the sugar
bowl."

Mary Jo: "Yes. Every meeting you went to."

Bruce: "The state and the utility was in it thick as soup, right
away."

Jim Nelson was troubled by what he saw at the Benson hearing:
"After Hartman read his little prepared statement, then he
winked at, I think it was Hancock, one of the guys from Commonwealth Associates, just sort of grinned at him. . . . Here the
state's supposed to be objective and independent; I always figured
they'd cave in to the power of the power company, but I didn't
expect this collusion to be quite that bad between the two of
them. And that was just sort of the beginning of the end of my
naiveté on that score."

Among the farmers, there were many such suspicions and
rumors and also many theories about why the hearings were
stacked against them. The Paulsons felt they were up against
strong presuppositions:

Mary Jo: "I think most of the people just thought we were a
bunch of kooks. That they genuinely believed that the power was
needed. And because they weren't experiencing it themselves . . .
you have to experience it yourselves before the full implications
come through. And I think that everybody at that level is committed to energy sources as they are now. Rather than developing
individual energy. . . . They were working with the assumption
that the power is absolutely needed for the majority of the people
in the state; then you can understand the way they went."

Bruce: "They didn't look into nothing. You had to bring it up
and then you were almost required to prove it. The state didn't do
nothing."

Jim Nelson also supposed there were personal dynamics: the

hearing officer, EQC staff, and power company lawyers "would come out in the country and every day they would face a new mass of angry farmers, and then after they got done facing all these angry people then they'd go back to their hotel—usually they would stay at the same hotel—they'd probably go and have a couple of drinks at the bar together. In any event, they'd all see each other in the restaurant and talk it over and I wonder how it's going to be tonight and just build a kind of camaraderie. It's them against the farmers."

That is how the farmers began to perceive the hearings. Nevertheless, large numbers showed up every night, many of them driving long distances after working since sunup in the fields. For some, it became a preoccupation. Uncompensated, they trailed the paid professionals all over western Minnesota. Virgil Fuchs went to every one of the hearings; he felt he had to. "It is such a wear-down kind of a thing. . . . People that didn't follow it, didn't really go and spend damn near their whole lifetime—all their days studying it and trying to keep up-to-date on it—didn't really have anything to talk about at the hearings. You really had to be involved in it to know because you would be speaking across the table from power company attorneys and whatever you said— they could immediately debate you."

Nevertheless, many farmers did speak. Jane Fuchs remembers the hearings this way: "They had their propaganda that they put into the record and then the people got up to say what they wanted to say." The farmers' statements ranged from the simple to the eloquent. They were united by a single underlying theme: the values of rural life and agricultural land. What Stearns County farmer George Glieden told Hearing Officer Evarts before a crowd of over 150 in Paynesville was typical of what Evarts heard all over western Minnesota:

> . . . Quite often you hear Americans boast what good Christians we are, good Christian-thinking people, and I say, if we are truly what we say we are, then I believe we have a responsibility and a moral obligation to the future and for the next generation. And I am referring to destroying good productive farm soil. And with these powerlines coming across our land, they will destroy hundreds of acres of good productive farm

soil that is so vitally needed for the production of food for the next generation. And we really don't know of the many hazards that a line of this type can create. Once these power towers are built, people will have to tolerate them perhaps for a good part of a century. And once good productive farm soil is destroyed, it is gone forever. I guess I could talk at some length on this yet, but I believe I have made my point. And I just have this yet to say. When these powerful power companies totally disregard some of the things I have mentioned, I believe this is morally wrong, and it is also my opinion it is evil. And all it takes for evil to prevail is when good men and women sit back and do nothing. I thank you.

One senses from the farmers' testimony that the powerline had already become a symbol—a symbol of the threat to a besieged rural "we" posed by the excesses of an urban and suburban "them." At the Glenwood hearing, Keep Towers Out (KTO) president Jim Clowes captured the spirit of this feeling:

With all due respect to the UPA and CPA, it seems to me like they have marked this transmission line out in the flattest and the best farm ground where it be the easiest to go across. Now, I must admit if I was going to go to Minneapolis, I would just as soon get on the highway as try to travel through hills and so forth, but, ladies and gentlemen, we have got to consider how much land we are taking out of production in the United States every year . . . we are losing two million acres a year to everything, everything that is being built, powerlines and roads and supermarkets and everything. Well, two million acres a year every twelve-and-a-half years adds up to a chunk of land the size of Minnesota, not just the farmland, the whole state. I think it is time we stopped doing this.

From this perspective, the powerline represented values the farmers could not accept. As Verlyn Marth pointed out, nowhere are these values more explicit than in the quantitative way Commonwealth Associates chose the route. Parks were excluded, wildlife areas avoided, and so forth, by assigning any square mile containing them a large avoidance number. Farmland didn't count. In Marth's words: "They assigned farmers zero. The computer routed it on farmers assigned zero only, and it did this in 1972 without any knowledge of anybody here. They sat around with their plastic suits and their white shoes around the table back in Michigan and destroyed these people's lives in secret two years before it was ever sprung out here where the route was going to be, without asking anybody, without ever seeing it, with no sense of values of the rural people or no human consideration on what they were doing."

This was no accident according to Marth. "They got by with these things a generation ago, so it looked like an easy way to go. That was a fundamental management mistake. It looked like an easy way to go and to them the reason of putting it in farmland is that they don't have to clear the land off to build this dumb thing, and that the farmer is going to maintain it under the line for the length of the line. That is why it's put on the prime farmlands in this state of Minnesota and that is the fundamental error of this thing. This thing is targeted for the farmers of the finest farmland in the whole world, I think. My grandfather homesteaded a hundred years ago here in Grant County. There never was a crop failure on that farm in three generations; that is the kind of land it is . . . they think that a farmer is at the bottom of the school well. The announcement is that that is not true anymore and the EQC better realize it and the power company people better realize it."

The kind of alternative values that *might* have been adopted in routing the line is suggested in the testimony of Marth and other witnesses, including Ben Grosz, a leader of Save Our Countryside (SOC).

Marth: Why do you have to avoid towns? . . . Why not put it right over all the towns. They're the ones that are getting the benefits. They're already destroyed. They're not nice farmlands. Why not put them right over the towns?
Grosz: . . . it is time for the people of the United States to stop being shielded or insulated from the responsibilities of their actions. As an example, if Minneapolis needs power, let them live with the smoke, sulfur, and haze associated with coal plants—not the people of North Dakota. If the state of Minnesota is serious about the environmental impact of trans-

mission lines, let the lines be built on existing right-of-ways. Granted an auto ride from Minneapolis to Fargo would not be a scenic one, but maybe a few people might start thinking about the real cost of this continued growth policy that is advocated year after year. As long as people are shielded from the results of their actions, no needed changes are ever going to be made.

The power companies did not even bother to respond at the hearings to this kind of testimony. They did not have to. It pertained to values, and the values did not fit those that government was responsive to at that time.

Health and Safety

One kind of testimony that could not be ignored was technical criticism—notably that the line might pose a threat to human health and safety.

By chance, there was a local expert on ozone production, Dr. Merle Hirsh, chairman of the science and mathematics division at the University of Minnesota's Morris campus, located right in the middle of the study area. He questioned the contention of Commonwealth Associates' expert, Hancock, that the concentration of ozone produced by the powerline would be well below that permitted by Minnesota law, contending that ozone production is very difficult to measure experimentally. His own theoretical calculations suggested a considerably higher ozone production rate from the powerline, one that is "dangerously close to the Minnesota maximum level." However, he noted in his testimony that his own conclusions were tentative and required experimental verification. The most he could say at that time was "the case for these pollutants to be below the danger threshold is still open to question."

Another health and safety concern was long-term exposure to low-level electric and magnetic fields. This is quite a controversial issue in the scientific literature, with some scientists, notably Dr. Andrew Marino and Dr. Robert Becker of the Veterans Administration Hospital in Syracuse, New York, claiming that results of experiments with rats exposed to AC fields suggest adverse health effects. The principal evidence claimed to suggest adverse effects on humans is in studies carried out in the Soviet Union on railroad employees working in proximity to AC power-lines.

This issue was brought up at the Elbow Lake hearing by two leaders of the farmers' protest, Kathleen Anderson and Jim Nelson. Anderson introduced as evidence correspondence with Dr. Marino, in which he described the results of his experiments with AC fields and asserted that very little was known about the effects of exposure to DC fields.

Nelson introduced an English translation of a Russian scientific paper in which it was stated that studies by another Soviet research group "showed that long-time work at 500 kv substations without protective measures results in shattering the dynamic state of the central nervous system, heart and blood vessel system, and in changing blood structure."

The reaction to Nelson's exhibit was quick and vehement. It came from an unexpected quarter—the EQC staff. Dr. Emmett Moore, chief of the entire EQC staff, immediately rose to cross-examine Nelson. One might have expected him to probe the substance of the Russian results, perhaps introducing additional evidence that the EQC itself had been able to find into the record. Instead he simply tried to have Nelson's exhibit stricken from the record:

MOORE: You have presented in evidence a Russian paper by V. P. Korobkova, *et al.* In that paper appears a statement quoting another paper; is that correct?

NELSON: Right.

MOORE: That statement which you quoted essentially referred to medical effects observed in another paper; is that correct?

NELSON: Right.

MOORE: In the paper by Korobkova there is no medical evidence, no study in that paper, just a quotation of another paper?

NELSON: My purpose in answering the—

MOORE: Would you answer me? Is that correct?

NELSON: That is correct.

MOORE: Now, do you have a copy of the paper by Asanova which apparently contains the medical evidence?

NELSON: I do not have a copy myself.

MOORE: Mr. Examiner, I request that you strike from the record any reference to the paper by Asanova unless Mr. Nelson can furnish a copy for us within a reasonable period of time.

NELSON: ... the power company has also referred to this paper in their material and I would request that perhaps it is easier for them to make a trip down to the Cities and provide this paper.

Evarts said that he would accept the paper, but unless he got a copy of the original study, he might "not give all the weight to the study that you'd like to have me give it." He contrasted that to weight he would give to the testimony he had heard previously by Hancock of Commonwealth Associates: "I would observe Dr. Hancock was court-tendered as an expert witness."

Later, John Drawz, CPA's attorney, renewed the attack on Nelson:

DRAWZ: Mr. Nelson, did you do a translation of either of the Russian papers?

NELSON: No, I didn't.

DRAWZ: Do you speak or write Russian?

NELSON: No.

DRAWZ: My next question is, do you either read or write Russian?

NELSON: No, I don't, and as I understand it—

DRAWZ: Thank you. You have answered my question.

At this point, many of the farmers in the audience evidently felt that Nelson was not being treated fairly. There was considerable booing and shouts of "Get out of there. . . ."

One person in the audience at that hearing was Owen Heiberg, editor of the *Herman Review*. He was appalled by what he saw: "You know, you don't know what's going on in the mind of that hearing officer; you don't know what's going on in the mind of those staff people from the Environmental Quality Board; but they, to me, the impression I got is they looked bothered about what these people were saying. And they always seemed ready to discount it, and very often on some technicality. It was very easy to sit there and get very impatient. You'd sit there and hear somebody, a farmer like Jim Nelson, stand up and he would make these statements, and you would wonder, 'Is what he's saying true?' Okay, that's what I wanted to know, 'Is what he's saying true,' and I'm sure that's what everybody else there wanted to know, too. But that question was rarely addressed. Most of the time when they would talk to Jim Nelson they would say, 'What are your credentials, what kind of credentials do you have to be making these statements?' Examining the kind of authority he was.

"Maybe that's appropriate when somebody is on trial for their life, but here we're trying to get at some kind of truth about an issue and he was making some statements, and really what you wanted to hear was somebody maybe on the other side making some other kinds of statements about this. Trying to get at, was what he was saying really true or not, and it got bogged down in this kind of morass of technicality. I remember sitting at one meeting and wanting—I didn't do it—but wanting to stand up and just say, 'Dammit, there is a moral issue here, and you're just treating it as if it is all just technicality.' And I think a lot of people who went to these meetings worked up the same kinds of feelings. I think it's almost impossible to sit there and not work them up."

That is where the health and safety issue stood in the corridor hearings. Expert testimony by Commonwealth Associates said there was no problem. One local scientist said he was not sure about ozone and two of the farmers had dug up documents suggesting possible problems from long-term exposure to low-level electric and magnetic fields. At the August 24 hearing, Evarts tried to deal with these concerns by reading a letter he had received from Warren Lawson, commissioner of health and a member of the EQC. Lawson said in his letter that the Minnesota Department of Health had launched a study of health and safety effects of high voltage transmission lines. As it turned out, however, that study was not completed until long after the state hearings on the powerline were over.

What Could They Say?

The major problem with the corridor hearings from the farmers' perspective was that they wanted no line at all and the point of the hearings was to determine where to put it, not whether it should be built. This led to continued admonitions from Evarts that arguments the farmers wanted to make were out of order. This happened, for example, when Verlyn Marth argued for a coal-fired plant near the Twin Cities as an alternative to the powerline:

EVARTS: . . . Mr. Marth, how are you going to get from a point in North Dakota to a point near the Twin Cities without traversing a certain amount of agricultural land.
MARTH: Mr. Evarts, . . . you're going to haul the coal down there and you are going to pour the pollution on top of them. That is what you are going to do.
EVARTS: We aren't contemplating that in this proceeding.

There were repeated attempts by the farmers to argue that the line should not be built because it was not needed. Evarts tried to head off discussion of need in his opening remarks before the hearings, as for example at Glenwood: "We aren't concerned here with the wisdom of constructing the line that I have described. That is beyond the scope of this particular hearing." At Glencoe, he read a letter from the director of the Energy Agency, a member of the EQC, saying that discussion of need in the corridor hearings would be "both immaterial and redundant" since the Energy Agency would hold hearings on that subject later. Nevertheless, it kept coming up. For example, at Elbow Lake, when Jim Nelson asserted that the powerline could be delayed until superconducting technology was developed to permit economic underground construction, Evarts ruled him out of order:

NELSON: . . . I'd like to include as an exhibit the power consumption curves of CPA and UPA as presented to the Minnesota Department of Agriculture . . . where it shows the past consumption. And now, this is my own conclusion, looking at them they generally level off the last couple of years, and my conclusion from that is that this powerline could be postponed until the mid-1980s when it could be undergrounded and be supercooled.

EVARTS: . . . I ruled previously that testimony or consideration with respect to need was not relevant for the purpose of this hearing. . . .

If their values were irrelevant, if they couldn't talk about need, if they didn't have the expertise to discuss health and safety, what was left for the farmers in these hearings? The answer is clear. What they could say was, *"Please put it on someone else."* That is what the corridor and route hearings boiled down to. That is the direction the farmers found themselves channeled in by the hearing process.

Thus, the hearings pitted farmer against farmer. If one farmer came to the hearing to say why the line should not go across his land, he was asked in effect to name what other farmer's land he wanted to put it on.

Some resisted. At the Paynesville hearing, Belgrade farmer Marvin Lensing refused to play the game:

LENSING: [We] worked all our lives to get this far and now we should give up the nice rich land right through the middle by diagonal crossing? No. It's not fair to us after all we've gone through to go right in there and chop a chunk out of the middle. That's what they want to do, chop a chunk out of the middle.
HARTMAN: Do you have a preference for a corridor in the area on that map?
LENSING: Less productive area.
HARTMAN: Would you point out where you think it should go?
LENSING: I will not point it out. That is for you to decide.

Others protested the constraints. At the first hearing in Olivia, CURE president Harold Hagen objected to what seemed to him to be an arbitrary limitation on the northern boundary of the study area before the public had a say. The ECQ's George Durfee responded:

DURFEE: Yes. It is an arbitrary limit.
HAGEN: But am I correct in assuming that even though it is arbitrary, it remains as a fixed boundary, immovable or non-adjustable?

EVARTS: . . . It cannot go north of that line. Does that answer your question?

HAGEN: It answers it in that you say it can't be. I don't see any reasoning, but then maybe I am not entitled to know the reason.

Still others questioned fixing the entry point from North Dakota into Minnesota prior to the hearings. When state representative Gene Wenstrom suggested at the Elbow Lake hearing, "To conduct public hearings when many of the major decisions are made seems to most people to be rather illogical," Hearing Officer Evarts replied firmly:

> Fortunately or unfortunately as it may be, we are locked into the situation where we can consider only the point of entry which is described in the application. At one of the earlier hearings a letter was made part of the record from the Public Service Commission of North Dakota indicating that there was approval of the routing in North Dakota, which of course determines or coincides with the point of entry we are talking about in the establishment of a corridor; . . . We have to take the application as it is and consider it on that basis and I would indicate for other witnesses this evening that the question of the termini of the line is not a matter that is subject to inquiry or dispute.

Ironically, Evarts' interpretation was wrong. What the letter from the North Dakota Public Service Commission actually said was, "We have learned that a substantial portion of the right-of-way from the point of crossing westerly has been acquired. Because of this fact, the Commission will be compelled to accept this route if and when they apply for certification under our act. . . ." The letter did *not* say North Dakota had already designated a route or that the Public Service Commission would refuse to grant a route entering Minnesota at another point. This left Minnesota with much more flexibility in routing the line than Evarts implied. This never came out clearly in the hearings, though perhaps the point was moot since the EQC had adopted a resolution fixing the entry point before the hearings began.

As a result, alternative corridors outside of the study area were not considered in the hearings. In particular, many of the original protesters wanted to discuss a proposal that the powerline run parallel to another line then being built from the same area of North Dakota into northern Minnesota. They suggested following the other powerline due east and then cutting directly south into the Twin Cities. Jim Clowes, president of KTO, did present this proposal at the Glenwood hearing, arguing that the land traversed by the alternative route would not be productive farmland. Because it was outside the study area, however, this proposal was simply disregarded.

With no real alternative, many of the farmers who appeared at the hearings explained why the line should not go over their land. Farmers in the southern corridor (II) said corridor I, the power companies' preferred corridor, would be better. One witness at Benson, in the heart of corridor II, told Evarts, "Towers necessary to support high voltage transmission lines would create a negative visual impact on the relatively flat agricultural land of the alternate corridor (II) as opposed to the more rollng topography of the other corridors." At Paynesville, in corridor I, there was general agreement that the northern corridor (IV) would be preferable. As one farmer put it, "I would recommend the line go through land with less potential value and the northern corridor would be my choice." A resolution of the Paynesville Jaycees recommended "a different route to the north in less productive land." When the hearings moved up north to Long Prairie in corridor IV, however, the farmers there disagreed. One witness pointed out that the farms in corridor IV were smaller; the powerline would do proportionally less damage to individual farmers if it ran over the larger farms to the south.

The farmers soon came to realize what kinds of arguments counted in these hearings and what kinds did not count. The one agricultural argument that seemed to make a difference was the potential interference of the powerline with irrigation. One group, the Bonanza Valley Irrigators Association, was well organized and well prepared to capitalize on this argument.

Another device effective in exorcising powerlines was an airport. According to FAA regulations the proposed powerline could be no closer than three miles from a sufficiently large airport. Suddenly some very small towns, like Starbuck, population 1138, and Brooten, population 615, began contemplating major expansions of their airport facilities.

The most famous was what later became known to powerline protesters as Starbuck's "paper airport." The mayor of Starbuck announced at the Glenwood corridor hearing that the city council was considering a hard-surfaced runway two-thirds of a mile long with lights and low-frequency radio for instrument landings. In light of the FAA regulations he just wanted to make sure the EQC knew about this possibility in planning the powerline. (Some years later, when we asked one powerline protester whatever happened to the proposed Starbuck airport, he replied sarcastically, "They built it in Dallas.")

A Citizens' Committee

While the hearings were going on, a citizens' committee, charged with making an independent evaluation of the corridors, was also at work. Based on the nominations of state legislators, county and regional commissioners, utilities, and environmental groups, the EQC named forty-seven citizens to a corridor evaluation committee. Among them were some powerline protesters, including Virgil Fuchs, Harold Hagen, Jim Clowes, and Kathleen Anderson, some officers of local utilities, including Charles Anderson, now president of CPA, and some county officials, including Michael Howe from Pope County.

The committee demonstrated right away that it would be independent. It headed off EQC's plan to designate a chairman by insisting on deciding its own leadership. Michael Howe recalls, "At the first meeting George Durfee and Larry Hartman told us that the board was going to appoint a chairman. And of course that all of a sudden hit everyone pretty sour. You know. Why can't we appoint our own chairman? A group of us got together and decided we're going to at least put some pressure on them to let us determine our own chairman." The EQC had little choice but to accept the committee's vote. Robert Braseth, county prosecutor from Traverse County, was elected chairman and CPA's Charles Anderson was elected vice-chairman.

The committee took its responsibilities seriously. An active core group of about thirty citizens worked steadily for three solid months investigating corridor options. The full committee met seven times. Subcommittees on electrical environmental effects, land use trends and patterns, agriculture, irrigation, transportation corridors, and natural resource features met separately and prepared recommendations.

Still, some committee members were uneasy about serving. As the first such group to function, they were not certain what influence would come of all their work and they feared being co-opted. This uneasiness was not assuaged by a letter they received near the beginning of their deliberations. The letter was from Charles Carson, a well-known environmentalist from Grant County, who had recently served as deputy director of the Minnesota Pollution Control Agency. It was such a powerful statement from such a respected, informed source that the entire committee discussed it at one of the early meetings.

Carson began by urging the committee not to neglect the question of need, even though that was formally outside its charge:

> I know that electricity is a wonderful thing; it has helped rural Minnesota a great deal. . . . *But* any good thing can be carried too far. . . . Stripmining Dakota and urbanizing Minnesota are not boons, and our descendants as well as ourselves will curse the day these things are done.

He asked them to think about what they had gotten into:

> . . . the EQC and outfits like it are stacked decks, trojan horses if you will, to absorb the energy of citizens, and neutralize their opposition to rip-offs, to make them think they are being honored with appointment to government, when, in reality, they are being used. Doesn't sound nice, does it? Well, don't take my word for it. How many votes do the 40 of you have? None. If there were 4000 of you, the result would be the same; 8 or 9 political appointees, of whatever party, will make the decision, not you, and that decision will give the power companies what they want, whether it is bad or not. You can advise until hell freezes over. . . .

The EQC was designed and set up to aid large corporations in getting around local government. The instant local government opposed UPA-CPA, for instance, they immediately ran to the EQC, even though they legally were grandfathered in under the new law so they wouldn't have to. During the Nixon years ideas like the EQC were taken out of Establishment think tanks in the East and put into practice. And that has happened in Minnesota. In the name of the environment, liberalism, and all that's holy, the people are being helped to dig their own graves. The EQC is destroying your local government, your own freedom.

What's to do? Well, I suppose most of you, ever hopeful that the politicians are helping you, will continue to play the game. . . . You can, in mass, challenge the EQC itself, hold a press conference and denounce its pandering to the power companies and demand that it be disbanded and government returned to the people. . . .

The committee was not ready for that. However, most of the members were from rural communities and most lived within the study area. This gave them a different perspective and different values from Commonwealth Associates and the EQC and led them to question the corridor preferred by the power companies.

From the beginning there was friction between the committee and the EQC staff. Michael Howe recalls some of the reasons. One was finding out about Larry Hartman's background: "The biggest fishy thing that I recall was Larry Hartman coming from Commonwealth Associates . . . and that was so glaring I couldn't believe it—I really couldn't believe it." Another was the trouble they had in getting help from the EQC staff: "Shortly after the first few corridor committee meetings, it was real obvious the EQC wasn't in any way trying to give us a fair shake on it. We asked for more technical information; we asked for detailed aerial photos; we asked for detailed soil maps and detailed irrigation maps . . . we didn't get it or we got it the day they wanted us to make a decision on it. . . . It seemed we got almost no help from the EQC people."

The committee resisted the constraints the EQC had placed on it in fixing the point of entry and limiting the study area. According to Howe, "One of the things we wanted to do was get that point of entry flexible, into Minnesota. I guess that was one of the first big decisions of the EQC that set the pace for that corridor committee.

. . . No leeway at all on it. That set the pace for a lot of it. What the hell's the purpose of having a corridor committee—a committee of people trying to set a corridor, if you don't give them some flexibility—we pretty much felt we were strapped with what guidelines they gave us."

Twenty-three members of the corridor committee, a majority of those active, petitioned the EQC, requesting that the study area be broadened. Behind the petition was a desire on the part of many of the signatories to recommend a corridor far to the north of the study area. When the EQC failed to respond to the petition, the committee was pushed even farther into what was fast becoming an adversarial relationship with the EQC staff.

An Adversarial Relationship

That relationship became evident on August 27, near the end of the corridor hearings. Larry Hartman took the stand at the public hearing in St. Paul and announced that the EQC staff had a corridor recommendation. He called it corridor V. It contained part of corridor I and part of corridor IV.

At great length Hartman explained the rationale for the recommendation, concluding that corridor V "best represents the interests of all involved parties. It is the belief of the Power Plant Siting Staff that this corridor reflects the testimony presented at the hearings. This corridor complies best with the EQC selection criteria. It is a corridor which represents and encompasses a variety of land use factors which present satisfactory routing possibilities. In conclusion, it is the belief of the staff that this corridor recommendation represents the most prudent alternative available."

There was one other feature of corridor V, which stood out to the farmers and the citizens' committee. It contained every inch of the route that UPA and CPA had proposed before submitting their application to the EQC.

After his presentation, Hartman was available for cross-examination. His principal interrogator was Bob Braseth, chairman of the corridor evaluation committee.

Braseth was incensed. He sought to show that there was no way such a corridor could result from a disinterested application of the

EQC selection criteria. For example, he questioned Hartman about
one particular area in Pope County near the town of Starbuck.
Right on the southern corridor boundary, but lying within the
corridor, was Glacial Lakes State Park. Immediately to the north
of the park ran the CPA-UPA preferred route.

> BRASETH: I refer you to MEQC Regulation 74, criteria for the
> HVTL corridor selection. They have exclusion criteria and
> they also have avoidance areas. Now, the first paragraph
> speaks of state parks as avoidance areas, does it not?
> HARTMAN: Yes.
> BRASETH: So, if that criteria had been applied in this locale,
> the line since it appears to be the border, would have been to
> the north end of the state park. Is that correct?
> HARTMAN: Yes.
>
> . . .
>
> BRASETH: Okay. So what I am saying here, is the fact that you
> have included it is not in complete agreement with the selec-
> tion criteria? In other words, had you moved two miles north
> approximately with the southern boundary of your corridor,
> of your recommended corridor, that park would have been
> excluded. Would it not?
> HARTMAN: Yes. Also on the other hand, you're not going to go
> along and put doughnut holes in a corridor to exclude every
> state park.

Braseth pressed his point by noting other features of that particular area:

> BRASETH: Is it a fair presentation to state that there are mu-
> nicipal boundaries involved, there is also an airport involved,
> there's also a state park involved, there are also conservancy
> lands and lands of a very fragile nature involved, there is also
> an area where the legislature is looking to purchase addi-
> tional lands involved. Is that correct?
> HARTMAN: I'm not sure about the legislature.
> BRASETH: But how about these other items?
> HARTMAN: Yes.

In conclusion, Braseth suggested, the only reason that partic-

ular area was included in the corridor was because it contained
the CPA-UPA preferred route:

> BRASETH: I wondered if that was not included because of the
> sole reason that that particular area was designated as an area
> which the applicants here had advised the EQC of at the time
> they filed their application, and if this was allowed in there
> just because they had it in there before.

What evidently disturbed Braseth was a suspicion that a deal
had been struck between the power companies and the EQC before
the hearings had begun and before his citizens' committee had set
to work. As he put it later in his cross-examination of Hartman,
"Apparently, some arrangement . . . was made with the EQC prior
to the designation of corridor I so that only a particular line could
be delineated." He hammered away at Hartman, pointing out
what he described as "inconsistencies" from one end of corridor v
to the other. He concluded, "It is my position that the various
criteria that the EQC has established by the rules and regulations
have not been met by the staff."

Verlyn Marth followed Braseth in cross-examining Hartman.
He suggested that Hartman's previous employment with Commonwealth Associates, the consulting firm that planned the
powerline, represented a conflict of interest:

> MARTH: Mr. Hartman, have you had a hand in picking this
> route?
> HARTMAN: In picking what route?
> MARTH: In this corridor v?
> HARTMAN: In the corridor, yes.
>
> . . .
>
> MARTH: Have you ever worked for Commonwealth Associ-
> ates?
> HARTMAN: Yes, I have.
> MARTH: Is Commonwealth Associates the group that sat
> back in Michigan and for a dollar fee took and assaulted these
> people with this proposed line in 1973, and are they the ones
> that are really responsible for all the troubles that are bent
> upon this woebegone line?
> HARTMAN: Well, that is a very eloquent statement. To re-

spond to that, I was not involved with that particular project. I was in Canada most of the time assaulting people with a +/−500 kv system.

. . .

MARTH: You don't think there is any conflict of interest between your past background and the fact that you are now picking the same corridor—

MS. HEILMAN: I object to this line of questioning.

. . .

EVARTS: I think I will sustain the objection Ms. Heilman made to the question because it was getting somewhat argumentative. There's no question but what Mr. Hartman has formerly been associated with Commonwealth Associates. Commonwealth Associates, however you regard the organization, is a collection of a group of highly technical consulting engineers that has engineered this phase of the study for the applicants . . . any public utility or any industry or member of an industry has occasion from time to time to go outside and engage private technical consultants.

. . .

EVARTS: Mr. Marth, wouldn't any group of consultants that came in for any utility raise in your mind the same degree of suspicion?

MARTH: Not when a man in a state agency and putting my tax dollars is prime in selecting the corridor that was—there is so much financial impact in this outfit for it and who he works for. How can we do that sort of thing? Maybe this man is honest as the day is long, but what you're getting is already people are so suspicious about this whole process. They have got five hundred people in that organization that got flared up and the farmers have never done anything like this before. The reason for that is the whole selection process is questioned. Now we find that a guy who is picking the final route works for these guys. How can we do that?

. . .

HARTMAN: When the project was in the office at Commonwealth Associates, I was in Canada. I was in Canada during the entire course of that project, plus I was seldom in the office. I have never billed one hour of my time to that project nor was I involved in it in any personal manner or fashion at all. Now I am employed by the state of Minnesota. My responsibility is to the state and to the citizens of Minnesota to help in delineating a corridor line route which best serves the interest of them.

But Marth persisted. Later he said:

MARTH: After all of this trouble, after two years we come up to the state of Minnesota, we are looking for a citizen protection, what do we get, somebody who works for Commonwealth Associates.
EVARTS: I'm going to rule that out of order.... Just forget it.

Norton Hatlie, CURE's attorney, continued this line of questioning, eventually obtaining the admission that the substance of the EQC "staff recommendation" was the judgment of just two people—Larry Hartman and George Durfee.

Braseth and many other members of the citizens' committee were upset with this recommendation. With only five dissenting votes, the committee decided to recommend an alternative corridor, which it called corridor VI. Located in the extreme northern part of the study area, corridor VI included a stretch of interstate highway I-94 almost from one end of the corridor to the other. It also differed from corridor V in another notable respect: it contained not one inch of the route UPA and CPA had proposed before submitting their application to the EQC.

The committee report describes all the effort that went into its deliberations. It contains a detailed summary of the testimony presented at the corridor hearings and a point-by-point discussion of how the committee's choice of corridor VI was influenced by information introduced in the testimony of many witnesses.

In discussing why it chose corridor VI, the citizens' committee, many of whose members were farmers, stressed the value of minimizing the impact on agriculture: "Consideration was given to the typical farm size, the crops grown, the present farming practices, the methods of irrigation or lack thereof, the soil conditions, the water supply which was readily available, the size of the equipment used in a given area, the economic condition of the communities adjacent to the agricultural area...." Of the corridors suggested, corridor VI "imposes the least impact upon the agricultural community and the agricultural economy in this state."

Having a citizens' committee was a departure from previous practice and the rhetoric in the citizens' committee report was a definite departure from that usually found in government documents. For example, the report bitterly criticized the EQC staff's corridor V recommendation, suggesting there may have been a deal:

> The purpose of the application may have simply been to circumvent various county ordinances which would have prevented the utility from utilizing their original route. The hazard in accepting this portion of Corridor V [the portion incorporating the original route] is that it leads to the speculation . . . that these entire proceedings have been had for the purpose of circumventing various county ordinances and laws and that some agreement was in fact made with the utility company guaranteeing to them that they would be given the corridor that they initially requested so they could utilize the routing.

That represented the feeling of most of the "public" who participated in the corridor selection process. Hearing Officer Evarts "weighed" all the evidence he had gathered and recommended what was essentially the Hartman-Durfee corridor. On October 3, 1975, the EQC adopted Evarts' recommendation and issued a certificate of corridor compatibility to UPA and CPA.

Two months later, at the first meeting of the citizens' committee assembled to help choose a route within the corridor, a sadder but wiser Michael Howe summed up what he had learned from his experience on the corridor committee: "If you want my honest opinion, I think it's a big front for the power companies, because the previous committee was given very few tools to work with; when they finally did come up with recommendations they quite strongly felt . . . they weren't listened to. Personally I think the whole committee is just set up as a rubber stamp for this process; I just can't see where there is any public input at all."

The Question of Need

Well, there's a couple of things that they better show

Like where is this current really gonna go?

Who makes the profits and who takes the blows?

How can you trust 'em if you just don't know?

from "Minnesota Line" by Nancy Abrams

After a corridor had been selected, the Energy Agency took up the question of whether or not the powerline should be built. The certificate of need process was really the key innovation in the package of institutional reforms. The power companies would have to convince the state that the powerline was really needed. According to the law that established the Energy Agency, among the criteria to be applied were "the relationship of the facility to overall state energy needs" and "possible alternatives for satisfying the energy demand." The farmers, ruled out of order in the corridor hearings when they questioned whether the line should be built at all, would at last have their chance. Or so they thought before the hearings began.

Although the Energy Agency had been established in April 1974, it was not until September 30, 1975, that the rules and regulations governing certificate of need applications were completed. They had been written by Richard Wallen, a bright, earnest young man with a Ph.D. in experimental physics and a master's degree in business administration.

Less than a week after the Energy Agency issued its rules, CPA and UPA applied for a certificate of need for the powerline. The CU line would be the first test of this new mechanism. Richard Wallen was assigned to handle the application for the agency.

Under the law John Millhone, director of the Energy Agency, had six months to decide whether to issue or deny a certificate of need. On October 14, the agency issued a press release, headlined HEARING SET FOR NEED FOR CONTROVERSIAL TRANSMISSION LINE, and invited anyone wanting to become a party to the proceedings to file "a petition for leave to intervene" within thirty days.

The need hearings were designed to be much more formal than

those held by the EQC. To participate fully, one had to be granted
intervenor status by the Energy Agency director. The notion was
that participation by the general public was less applicable to the
objective process of determining need; in the view of the Energy
Agency, need was a technical subject to be handled by technically
qualified individuals.

A rift developed among the powerline protesters over who
would represent them. Members of the Douglas and Grant County protest groups had not been happy with Norton Hatlie, the
lawyer CURE had hired for the corridor hearings. They decided to
find another lawyer. As a result, CURE was represented at the need
hearings by Hatlie and Grant Merritt, former director of the Minnesota Pollution Control Agency and a leading environmentalist,
while Save Our Countryside, Preserve Grant County, and No
Powerline were represented by George Duranske, an attorney
from Bemidji.

Lawyers Get Together

The difference between the need and the corridor hearings is well
illustrated by the first session. On December 1, hearing officer
William Mullin, a Minneapolis attorney, called the lawyers together for a prehearing conference. Present were Mullin, John
Drawz for CPA, Roger Miller for UPA, Norton Hatlie for CURE,
Richard Wallen for the Energy Agency, and Dwight Wagenius for
the state attorney general's office. The whole session was very
civil, very chummy—a group of professionals sitting down to sort
things out.

They talked about scheduling and procedures; they also talked
about substance, particularly, what issues would be appropriate
to the determination of need and what would be off limits. For
example, the power company lawyers insisted that a power plant
in Minnesota should not be considered as a possible alternative to
the powerline for satisfying the energy demand. According to
John Drawz, that was outside the scope of a process to determine
need:

> DRAWZ: We submit that this proceeding is not a proceeding
> where that is an issue: we are here to determine whether or

not there is a need; that is, whether there is an existing increased demand for electric power which will be brought in by the proposed facility. We are not here to discuss and there is not available as an issue, the location of that power plant under these rules and under the statute.

UPA's Roger Miller explained that the project was just too far along in December 1975 to consider turning back:

MILLER: . . . what we are saying is that we are well down the line, three years down the line, in the construction of the electric generating plant. The costs of that facility and the costs of the associated transmission facilities are being financed through the Rural Electrification Administration and the Federal Financing Bank, all of whom have approved financing the project. . . .

Later he added, "At this stage in our process . . . we are committed to a lignite-fired generator. The agency regulations weren't in effect until a couple of months ago. . . ."

Miller also contended that the EQC corridor selection had limited the options open for consideration in the need process: "It's been determined already by the Environmental Quality Council that that would come in by a direct current line and the terminal point of that line has already been determined, physical characteristics of the line have been determined. . . . So we are found in a much narrower band of alternatives than you would be starting from day one."

Mullin asked Wallen what the Energy Agency's position was concerning the alternatives appropriate for consideration in the hearings. It would take a long time to build a facility in Minnesota, said Wallen, "if you take the timing of need and of supply, as represented in the application, if that's correct, I'm not stating it is correct, but if you take that timing, the number of alternatives is extremely small. . . ."

That is a good example of the passive stance assumed by the Energy Agency in these hearings. Wallen might have told Mullin whether he felt the need and supply projections in the power companies' application were correct or not. Instead he did not commit himself. Later in the session, it became clear what he saw as the role of the Energy Agency in the hearings:

I would like to state what I see my role as. The Energy Agency has a lot of accumulated facts and figures. We have knowledge as to all utilities in the state . . . and it's my intention that the record be accurate, as accurate as possible, and at any time statements that are made that are clearly inconsistent with known facts, I would want to cross-examine that witness and ask him if he is aware of this particular document, the organization, and if PC Docket such and such reported such and such. That's information inconsistent with what you just stated. Could you clarify this inconsistency?

Evidently the agency planned to sit back, watch the power companies' and farmers' lawyers fight it out, and see what happened. In explaining this position to Mullin, however, Wallen inadvertently let his predisposition show:

WALLEN: At this point the agency does not intend to say whether this application should be denied or accepted.
MULLIN: As of December 1?
WALLEN: As of December 1. In other words we are willing to listen to Mr. Hatlie. He might bring some facts to our attention that we do not know of.
MULLIN: Well, you haven't heard what Mr. Miller's and Mr. Drawz' witnesses will say either. They may or may not be persuasive to you apart from their impact on me.

Another topic at the prehearing conference was planning for one session where the general public would be invited to testify. The Energy Agency proposed to devote one day to a public hearing and Director Millhone suggested that it be held in the heart of the protest area, in Alexandria. This was to be the opportunity for individual farmers to have their say.

One gets insight into how seriously this exercise in public participation was taken from the conversation as these professionals planned for that session and for other possible appearances by members of the public:

HATLIE: . . . I have been engaged for a number of years in cases of this kind. I would suggest that one of the ways that the proceeding can be expedited would be to amend the notice which binds our hearings in Alexandria . . . to show that public witnesses of the kind who expressed opinions only . . . be instructed to appear at 4:30 so that we lawyers can get at our business during the hours that are fixed. . . . I have found from my experience that an excellent way to dispose of this kind of appearance. And may I say during the corridor hearing my orderly cross-examination and arrangement of my witnesses were interupted when the hearing examiner would permit public harangue at 10:30 in the morning. I think there was a reason for public opinion evidence in that case, but in this one it appears that we are getting at the hard stuff.

MULLIN: . . . it's my reaction that if someone comes to the hearing and he is not represented by a lawyer and he really has got something that he wants to get off his chest, it's more in keeping with the spirit of the Act to interrupt the lawyers and witnesses who, after all, are getting paid for being there, and allow this fellow to speak his piece. I don't think it's very fair to pass a regulation that allows people to make statements and then, you know, keep them cooling their heels for three or four days waiting for an opportunity to say something.

DRAWZ: I would respectfully disagree with you. . . . I would ask you to just compare the Energy Agency Act and the Power Plant Siting Act . . . [N]either the statute nor the rules concerning the Power Plant Siting Act contemplate formal parties or intervenors. . . . So I would say that this is a very narrow technical confined proceeding limited very much to parties and that if citizens have something they wish to say, that they would come, as Mr. Hatlie suggests, at a time when it would not disrupt the main proceeding. . . .

WAGENIUS (attorney general's office): In section 116H.13 on certificate of need I find only two references to public hearing and that's all it says. . . . "In reviewing each application, the Director shall hold at least one public hearing pursuant to Chapter 15." There is no expressed legislative intent that a full-scale, you know, public participation, hearing be held, but rather a public hearing pursuant to chapter fifteen, which is, as we know, more formal.

MULLIN: . . . I guess you are all against me on this a little bit. You did pass regulations that said that people could come in

and issue statements, and I am still very reluctant to, you
know, in effect read that regulation out of the definition of
public hearing by making it so inconvenient for people to participate in the public hearing that they are going to be sorry
that they ever got involved. . . .

. . .

HATLIE: . . . when I, as a paid lawyer, . . . have to orches-
trate a horde of people in the middle of my business, it
becomes rather a burden. . . . [Y]ou are certainly going to
get a great deal of opinion evidence that will stray far beyond
the issues which will control us all in the fact-finding procedure.

Mullin then read the Energy Agency regulations and concluded
that he was mistaken: "This reading, it seems to me, gives rather
powerful support to Mr. Drawz' position that the intent here of
this proceeding is to be a much more orderly presentation."

They decided to limit the public participation to Alexandria on
December 15. Later on, in discussing what issues are relevant to
need, they agreed that environmental effects are not to be considered. Hatlie, the farmers' lawyer, also contended that health
effects should not be an issue. But then Mullin said he felt uncom-
fortable about ruling things out in private and then going out and
letting people talk about them at Alexandria: "Don't you kind of
get the feeling of . . . well, you know, we will all be winking at each
other when people want to come in and present testimony on the
Russian study, or something that may have to do with health
effects and we all know that health effects have nothing to do
with this proceeding."

UPA attorney Miller said he thought they had made real progress in the prehearing conference in defining the issues so that the
hearing would not be "clouded in irrelevant testimony." He described the corridor hearing as "close to nineteen hundred pages
of repetition."

CURE's Norton Hatlie supported this view, giving, as one example: "Marth appeared three times at three different hearings for
at least one hour at each hearing."

Miller added, "And said the same thing each time. . . ." Later he
characterized the corridor hearings as "an unbelievable record."

HATLIE: Those people are going to show up and they are going to have their say and if my estimation of what happened during the corridor hearing is correct, there is not much you can do to deny them once they bring their bodies into the hearing room. The best you can do is control them. I thought it got almost dangerous at one point when Mr. Evarts tried to impose the rule of law, as he announced it, on some of the people that wished to speak, and there was booing and there was hissing and clapping; so I guess that's why I recommended the setting aside of a portion of a day, or every day, so that they can purge themselves. . . . I can only say for the CURE people that aside from a few stragglers, which I cannot control, such as Mr. Marth, none of them will show up in the public portion. . . . Now I cannot control certain of the mavericks who insist on using the opportunity to be a lawyer . . . with Roger, I can't agree more. I think what we are here to do is lawyer's work, but I don't think you can isolate yourselves. I think you are going to be confronted early with a witness from the public sector who wants to harangue you and I don't know how you are going to handle that.

. . .

MILLER: . . . it's kind of like a hockey game. If you don't keep some control, it will be totally out of hand.

HATLIE: That's right. And I suspect in this situation you are going to run straight into that."

MULLIN: I see—I think the fact that we are having the hearing in the courtroom will have some effect in helping people to keep their decorum. Mr. Miller looks skeptical.

MILLER: We had some hearings in the courtroom on the last one.

HATLIE: Get a big gavel.

MULLIN: Maybe I could come down through the skylight.

"Expert" Witnesses

The formal hearings began on December 8 at the Hennepin County government center in downtown Minneapolis. From the beginning it was clear they would be a mismatch. The power companies had so many resources to call on—smart, seasoned lawyers, well versed in the utility business; well-prepared, knowledgeable witnesses, who had been working full-time for years planning the CU project; and unlimited access to information about the project and about the complex interrelations between CPA and UPA and the other utilities in the electric power grid. By contrast, the farmers' lawyers were not as experienced, informed, or effective, they did not have access to a pool of comparably knowledgeable witnesses, and much of the information they needed was in the hands of the power companies. The Energy Agency might have been able to redress this imbalance of personnel and information somewhat, but it did not choose to assume that responsibility.

First to testify were five expert witnesses on behalf of the power companies. In their minds there was no question about the need for the powerline. James L. Herbert, manager of CPA's engineering and planning division, who helped to prepare the certificate of need application, told hearing examiner Mullin:

> CPA will not be able to serve its present and future loads if this application is denied or delayed. There is no alternate feasible way to meet the demand. Therefore, it is my conclusion that the proposed facilities are of the proper size and type and are timely; they are in fact urgently needed. . . . In my judgment the evidence will show beyond a shadow of a doubt that the probable result of denial of this application will be an unacceptable level of reliability of electric service to electric consumers in Minnesota.

David Kopecky, assistant to UPA's general manager, had also worked on the certificate of need application. He summed up the UPA position:

> After considering all of the historical data and the forecast results . . . it is my conclusion that UPA has an urgent requirement for its share of the generating capacity from the Coal Creek station units. UPA already has a substantial deficit of generating capacity, which will be 153 mw this coming winter. We are considering a relatively short-term period from

now until the generating units should be completed. Even
if there would be some deviation in the load growth esti-
mates, this would not materially affect UPA's requirement
for this generating capacity. A review of the alternate assumptions that would be made in load forecasting indicates
that the load forecasts may be conservative rather than
too high. Therefore, it is my conclusion that [failure to build]
these facilities would [result in] an unacceptable level of
reliability of electric service to ultimate consumers of the
member distribution cooperatives of UPA.

The argument they presented was simple, straightforward,
and direct. The *need* of a utility for new facilities can be determined, they asserted, by comparing its *anticipated demand*
with its *assured supply:*

need = anticipated demand − assured supply

If CPA and UPA could demonstrate that their anticipated demand exceeded their assured supply, then their need to build
additional capacity to make up the deficit would be established.

According to the witnesses, the electric power supply available to the two power companies was easily determined. UPA's
generating capacity was 237 mw. 166 mw came from its Stanton Plant in North Dakota. Another 71 mw came from oil-fired
plants in Minnesota. CPA's generating capacity was 181 mw,
170 mw from its share of the output of a coal-fired plant in
Wisconsin. The other source of power for CPA was the United
States Bureau of Reclamation (USBR) dams on the Missouri River,
but it could only be assured of 111 mw from that source after
1976.

The combined assured supply for CPA and UPA reported at the
hearings was therefore:

assured supply = 237 mw + 181 mw + 111 mw = 529 mw

What about the anticipated demand? CPA and UPA witnesses
painted a gloomy picture of upward spiraling demand. They presented figures showing the "demand obligation" of the power
companies:

year	demand obligation
1974	779 mw
1978	1186 mw
1979	1316 mw
1980	1437 mw
1981	1571 mw
1985	2256 mw
1990	3600 mw

"Demand obligation," defined as the peak amount of power consumed during the year plus a 15% "reserve margin," is thought to be a sensible figure to match against the supply because the supply has to meet the peak demand. The reserve margin is included because some of the time generating equipment is down, due to malfunction or maintenance.

Of course, for the years subsequent to 1974 these figures were speculations. They were projections, based on the demand experienced in previous years and on certain assumptions about the future. CPA's Herbert spelled out those assumptions at the hearings. The most important one was that the demand would continue to grow exponentially at the same rate it had in the past, doubling every seven-and-a-half years.

Subtracting the assured supply figures from the anticipated demand (demand obligation), one finds what was purported to be the need:

year	need
1974	250 mw
1978	657 mw
1979	787 mw
1980	908 mw
1981	1042 mw
1985	1727 mw
1990	3071 mw

There was no way to meet this need, the company experts claimed, except by building the powerline.

If the companies' projections were correct and if the powerline and power plant were completed on schedule, with 500 mw coming on line in 1978 and another 500 mw in 1979, the revised need table would be:

year	need
1974	250 mw
1978	157 mw
1979	−213 mw
1980	−92 mw
1981	42 mw
1985	727 mw
1990	2071 mw

This need would be met between 1975 and 1978 by purchases from the USBR and the MAPP pool. In 1979 and 1980, the companies would have some excess power to sell. However, in 1981 they would once again be in a deficit situation, and by 1990 two more CU projects would be required to meet the need.

The other principal argument advanced by the power companies in the need hearings was that the project was simply too far along to consider turning back. Their lead-off witness, Reynold S. Rahko of UPA, the project coordinator for the powerline, presented a detailed chronology of progress on the CU project, including:

JUNE 1972—CPA and UPA, together with representatives from the REA met in June 1972 to discuss the feasibility of jointly providing for the future power and energy requirements of both organizations.

NOVEMBER 1972 — Burns and McDonnell of Kansas City, Missouri, was selected to make the feasibility study.

JUNE 1973 — The MARCA (Mid-continent Area Reliability Coordination Agreement) Council approved the CU project transmission proposal.

JULY 1973 — Burns and McDonnell completed the feasibility study for the CU project, along with an environmental analysis of the plant.

SEPTEMBER 1973 — Commonwealth Associates, under subcontract to Burns and McDonnell, completed the environmental analysis for the powerline.

NOVEMBER 1973 — The formal loan application was submitted to the REA.

NOVEMBER 1973 — Black and Veatch of Kansas City, Missouri, was selected to be the consulting engineer for the design and construction management of the CU transmission system.

FEBRUARY 1974 — The CU loan application was approved by the REA.

MARCH 1974 — Contract for two turbine generators was awarded, $28,700,000; contract for two steam generators (boilers) was awarded, $76,000,000.

MAY 1974 — Contract for HVDC terminals was awarded, $54,000,000.

JUNE 1974 — Site for Dickenson substation was acquired.

JULY 1974 — Site for Coal Creek generating station was acquired; water permit for Coal Creek was granted by the North Dakota State Water Commission.

SEPTEMBER 1974 — REA designated the Federal Financing Bank as the source of financing to supplement direct REA loans; contract for powerline conductor wire was awarded, $15,250,000.

OCTOBER 1974 — Coal agreement was signed with North American Coal, resulting in the creation of Falkirk Mining Company; contract for site clearing and earthwork was awarded, $1,200,000.

FEBRUARY 1975 — Contract for power station structural steel was awarded, $23,800,000.

APRIL 1975 — Application was submitted to the EQC for corridor designation.

MAY 1975 — Construction started at the power station; contracts for steel towers were awarded, $12,400,000.

Rahko concluded by stating that as of September 30, 1975, when the rules and regulations of the Energy Agency were completed, the status of the CU project was as follows:

a. Power station:
   engineering completed: 48%
   number of contracts awarded: 58
   amount of work under contract: $203,000,000

b. Transmission system:
   engineering completed: 44%
   number of contracts awarded: 17
   amount of work under contract: $90,000,000.

What is equally clear from the chronology is that as of March 1974, when the power companies found themselves in trouble in Pope County and decided to request the state process, very little engineering had been completed on either the power station or the transmission system and no contracts had been awarded. At that point CPA and UPA very likely knew they would have to undergo a certificate of need process; indeed, in early April 1974, when the EQC accepted the CU application, it explicitly stated that the question of need would be addressed later by the Energy Agency. Nevertheless, between March 1974 and September 1975, CPA and UPA proceeded full speed ahead with construction of the power station and powerline. As a consequence, by the time of the need hearings, they could claim the project was virtually a *fait accompli*. Roger Miller, UPA's attorney, explained to Hearing Officer Mullin, "You have a midstream situation, if the regulations became effective September 30th of this year, of a project which has been ongoing for three years. We are not starting from square one. . . ."

What About MAPP?

The intervenors groped for an effective tack to take in these hearings. Was it really too late to push for alternatives to the North Dakota power plant? Were there holes in the companies' projections of demand? Without expert witnesses of their own, they had to make their points in questioning the power company witnesses. The result was a scattergun barrage of objections, not a coherent challenge to the companies' assertion of need.

Norton Hatlie, representing CURE, chose to concentrate on alternatives to the CU project and on the notion that the power was not all going to farmers. George Duranske, representing the Grant and Douglas County protest groups, and John Gostovich, representing MPIRG, focused primarily on the validity of the demand projections, particularly on the potential impact of energy conservation, which the companies had ignored.

Hatlie's most intriguing approach had to do with the fact that
CPA and UPA were part of a large, highly integrated electric utility
grid. They were members of MAPP, the Mid-continent Area Power
Pool, and MARCA, the Mid-continent Area Reliability Coordination Agreement. They had entered into a complex array of agreements and contracts with other utilities to use the output of each
others' generating facilities and to wheel each others' power in
the network of transmission lines.

For example, although UPA built the Stanton plant in North
Dakota, witness Kopecky explained at the hearings that none of
its power goes directly to UPA customers:

> KOPECKY: The output of the Stanton plant reaches our con-
> sumers, or our system, through a displacement agreement
> arrangement that we have with Northern States Power Com-
> pany and Otter Tail Power Company.
> Q: What is a displacement agreement?
> KOPECKY: The displacement agreement basically provides
> that the power from the Stanton plant is transmitted to loads
> of the companies in North Dakota where they have con-
> sumers and the companies in return give us an equal amount
> of power from generating facilities in their system and de-
> liver that to our Minnesota system.

The planning for new generating plants and powerlines is done
by committees of MAPP and has to be okayed by committees of
MARCA. This insures coordination and reliability. It means that
the long-term planning for electrical development is done on a
comprehensive, regional basis by a consortium of utilities.

For example, the CU project did not really begin with the Burns
and McDonnell study commissioned by CPA and UPA in November 1972. It began several years before with a MAPP study. According to the *MAPP Newsletter* of December 15, 1969:

> . . . several MAPP members will jointly finance a study of the
> cost of linking two million kilowatts of generating capacity
> in the lignite fields of North Dakota with the load centers of
> the participating utilities including the metropolitan areas of
> Eastern Minnesota. A.W. Benkusky, MAPP manager made the
> announcement and listed the participants in the study as fol-
> lows: Cooperative Power Association, Minnesota Power and

> Light, Minnkota Power Cooperative, Montana-Dakota Utilities, Northern States Power Company, Northwestern Public Service and Otter Tail Power Company.... The study will be performed by Commonwealth Associates Inc. of Jackson, Michigan, and is expected to be completed by about February 1, 1970.

As another example, before the CU powerline could be built, it had to be approved by the MARCA Council.

Within MAPP and MARCA, CPA and UPA are like children among giants such as Northern States Power Company. During the winter of 1977, for example, MAPP as a whole experienced a peak demand of about 15,500 mw with NSP accounting for almost 4000 mw, while CPA had about 500 mw and UPA about 400 mw. At that time NSP owned generating facilities capable of delivering 6600 mw, while UPA and CPA's combined owned capacity was only about 250 mw. Since voting in MAPP is based on megawatts of demand, UPA and CPA have very little to say in the ultimate decision of that organization.

This situation gives rise to several questions about the CU project. For example, are the long-term, low-interest loans arranged by the Rural Electrification Administration going to the benefit of rural electric cooperatives exclusively or are some of the benefits going to investor-owned utilities like NSP? (The fact that the power delivered by the CU powerline is immediately transferred to NSP substations does little to assuage such suspicions.) More germane to the need hearings is the question of whether, for some time at least, MAPP utilities, like NSP, might have supplied power to CPA and UPA as an alternative to the CU project. More fundamentally, should the determination of whether the CU project was needed have focused just on the need of CPA and UPA, or should it have focused instead on the need of the entire integrated network of MAPP? After all it is MAPP that plans for such projects. If the need of individual utilities like CPA and UPA are the sole focus of need hearings, it is like playing a shell game with MAPP controlling the shells.

For instance, one strong argument used in the hearings in support of the need for the powerline was that one MAPP member, the USBR, had informed another, CPA, that after 1976, the power it could count on purchasing from USBR would be cut back. As CPA's

Herbert pointed out at the hearings, this significantly affected CPA's supply picture: "CPA will lose 214 mw of USBR-supplied capacity in 1977–78. This is approximately equal to CPA's share of Unit #1 of the CU project and is a significant part of the total needs set out in this certificate of need."

Early in the proceedings, Norton Hatlie proposed that MAPP itself be made a party to the need hearings, because "the very facility which we are considering today was planned from the beginning prior to 1972 by the applicants in connection with the MAPP organization and whereby each of the members of MAPP were entitled to one vote for each twenty-five megawatts of power consumed. . . ."

This proposal was immediately shot down. UPA attorney Roger Miller derided it, saying, "They intend to shove the kitchen sink into the proceeding to delay and resist. I cannot see any relevancy whatsoever in the line of examination of Mr. Hatlie to the proceeding before us. . . ." When Mullin asked for the Energy Agency's opinion on this issue, Dwight Wagenius backed up Miller. "Well, I guess the agency has problems in seeing the relevance of the line of questioning."

After considerable discussion, Mullin finally ruled that the Energy Agency did not have the authority to make MAPP a party to the proceedings. However, "the agency does have subpoena power that might reach into MAPP's files, if you can make a showing that . . . MAPP has some documents that these applicants don't have that might go to the [need] issue. . . ."

This, however, was a Catch-22 situation. The intervenors could see a particular MAPP document only if they could show it was relevant, but without seeing the document they could not make such a showing. Hatlie tried a broad request of CPA and UPA documents, including minutes of meetings back to 1970, in order to probe the MAPP connection. The companies refused, citing an Energy Agency rule stating that the right of discovery requires the showing of good cause. In the end all Hatlie got was a copy of the MAPP agreement, a document that he already had before the hearings began.

Paul R. Heim, general manager of the MAPP coordination center, located in Minneapolis, did appear as the power companies' final witness. He assured Mullin that, "The pool is not and is not intended to be a source of power for long-term power supply of any systems." In answer to the question, "Are new units being planned which could satisfy the needs of CU?," Heim replied, "As a short answer to the question, no, there are no other units being planned in the pool which could satisfy CU requirements."

This is as far as the MAPP issue was pursued at the need hearings. Without access to internal documents, and without the assistance of experts with an insider's understanding of its workings, MAPP remained a black box; the intervenors were unable to penetrate its surface.

Projecting Demand

The other principal thrust of the intervenors' inquiries dealt with the validity of the companies' projections of electricity demand. The time of the hearings, December 1975, was one of considerable uncertainty in the utility industry. The reverberations from the Arab oil embargo two years before had shaken historical consumption patterns and rendered problematic whether time-honored assumptions about electricity demand were still valid. As Heim put it, "The industry is currently in a precarious position with regard to load forecasting and capacity commitment. The historic pattern of load growth and analysis has been affected by the new conservation ethic, by the new price levels associated with shortages of oil and gas, by the change in daylight savings, and by the depressed economy."

The power companies' demand projections were predicated on the assumption that the exponential increase in consumption experienced in the previous ten to fifteen years would continue at the same rate. Indeed, their witnesses argued, this may have been conservative. Given the increasing price and uncertain status of oil and natural gas, more and more new homes were being heated with electricity and some older homes were being retrofitted with electric heating. If this substitution continued, they suggested, then the demand would outstrip the companies' projections.

On the other hand, intervenors John Gostovich and George Duranske noted, the projections did not include any possibility for energy conservation, nor did they include substitution of solar or other alternative energy sources that might lead to a demand

below the historical projections. Gostovich pointed out several
issues the hearings had neglected: "We have not addressed any of
the determinants for electrical consumption to date in this hearing in terms of its price, in terms of what it can be used for, in
terms of new technologies that may or may not come on line. . . ."

Duranske suggested that the hearing examiner should carefully
consider just what "need" is:

> Need is not necessarily equated with how much electricity
> we can sell and it is our hope to show that the amount of elec-
> tricity that has been sold in past years, which [is] the basis for
> the forecasting, has been artificially created . . . there was a
> saturation in advertising for the sale of electrical appliances
> and consequently the creation of a level of use of electricity
> which is not the same thing as a need for electrical power . . .
> if there is by virtue of the installation [of] the CU project,
> [a] substantial increase in the cost of electricity, that may also
> corrupt the need statistics and forecasting in the short run as
> well as in the long run.

The intervenors scored some points. For instance, in his cross-
examination of CPA's Herbert, Duranske brought out that the
companies' projections did not include any allowance for price
elasticity, that is, reduction in demand due to increased price of
electricity.

However, the companies' lawyers argued that to establish need
all they had to show was that their anticipated demand exceeded
their assured supply in the very near term. In an exchange with
Gostovich over the validity of the companies' demand forecasts,
CPA's Drawz made this point: "I have serious doubts as to whether Mr. Gostovich is even going to question the accuracy of our
forecasts as they relate to '78 or '79. I . . . believe he won't find
much wrong with them, if anything . . . a lot of the things we are
talking about may have much more meaning if we were talking
about a need for 1990, but this is a very short range we are talking
about." To Hearing Officer Mullin, this was a telling argument.
"Good point, " he said.

For their part, the intervenors were unable to present a specific
alternative program of conservation and substitution, with quantitative predictions of reduction in demand, that would counter

the companies' claim that the CU project was the only answer. For example, they never seriously explored the combination of buying power from MAPP in the short term and reducing consumption through conservation and substitution in the long term. They did not have the resources to develop and analyze an alternative plan.

The Public Hearing

The one public hearing, at Alexandria on December 15, was nothing like what Hearing Officer Mullin had been led to expect by the lawyers at the prehearing conference. It was impassioned, but it was orderly; it featured thoughtful, heart-felt statements about whether the line was needed, by people who would have to sacrifice if it were built. They homed in on the issue of conservation.

Jim Nelson of Grant County was the first to testify. He presented examples of CPA literature promoting increased electricity use. He noted that his own cooperative had a rate schedule that charged less per kilowatt-hour the more electricity was used: "Now this would be an interesting type of thing to use in a filling station, for instance. The more gas you buy, the less you pay . . . now some of the utilities tell us they are pressing for conservation of electricity, but still continue to use this schedule."

He concluded his statement by saying, "The best solution would be no line at all. If the power companies could quit pushing usage of power and start pushing conservation of power, then they could have a little extra time off looking into alternatives."

Ben Grosz of Douglas County, representing Save Our Countryside, continued the conservation theme:

We are convinced that the applicants' projected forecast for increased electrical need is based on their continuing with a high level of advertising and a rate schedule that encourages wasteful electrical use. We will continue to oppose any new transmission line construction across agricultural ground as long as a deliberate attempt is being made by the electrical industry, through advertising, to increase electrical consumption. . . . We would suggest to the hearing officer that he refuse to issue a certificate of need and, in turn, tell the two applicants to go home and stop encouraging electrical use and waste.

Grosz pleaded for consideration of the rural environment in deciding whether the North Dakota power plant and powerline system was the appropriate way to proceed:

I don't care if the applicants say the North Dakota generation plant will meet all federal and state pollution regulations. I've seen with my own eyes what the Black Mesa plant has done to the Arizona-New Mexico skies and it is an unbelievable crime against man and nature. And the plans are for twenty to thirty of these plants in the states of Montana, Wyoming, and North Dakota. We fail to understand what good electricity is in your home if you cannot breath the air. The utilization of coal to generate electricity is a dirty, air-polluting process and it will cost large sums of money to keep smokestack pollution to a liveable level. If these plants are permitted to be built out in the wide open spaces, air pollution will grow worse and worse. We suggest the generation plants be built at the center of need—in this case Minneapolis-St. Paul. . . . Isn't that a fine bit of hypocrisy? The coal plant is too dirty for the metropolitan area of Minnesota, but just right for the farmers of North Dakota. We say, don't build one more line or coal plant until the pollution problems have been taken care of so a plant could be located in a metropolitan area.

Noel Kjesbo, a young Grant County farmer, urged the rural electric cooperatives and the Energy Agency to reassess their missions:

Rural coops have a unique opportunity to relieve the so-called burden of projected usage of power . . . they have a rather unique opportunity to begin doing and encouraging things like wind power and solar power. I haven't heard of anything like that done. . . . I think in years gone by, the public utilities, quote, unquote, were set up to do a very important need, and that was to [serve] us with electricity. Now I think that . . . we can't go on growing forever. I think now at this stage of our life, the public utilities, maybe as well as the Energy Agency or the state, has an obligation to change its charter. Instead of serving us as you did by turning on power and whatever, I think you now have an obligation to help us conserve these natural resources that almost certainly face extinction.

Others told Mullin in simple, direct ways what they felt about the powerline. Francis Thompson, an elderly farmer from western Douglas County, arose to say:

I don't represent anyone in particular. I just came here to the hearing to hear what was going on. . . . I registered and marked the card that I would testify.

The only thing I have to say as far as testimony is concerned, I don't think that we have used any increasing amount in the last two or three years. I don't anticipate that we will use any increased amount in the next two or three years. Our farm happens to be in what was the proposed corridor. I don't know if it is in the present line of thinking for building, but if it is, I would never sign an easement for construction; I say that because I am getting to the age where I won't farm too much longer, but I have sons who probably will. I could never sign an easement permitting some power company to run a line across the farm that will almost ruin it for them.

Ben Herickoff of Stearns County, whose sons had recently taken responsibility for the land he had farmed, stated bluntly, "I never will sign an easement for this line. . . . I will fight, even if I am old. This line is not going to go through our land. . . . We took our place over from our ancestors in a real fine place, shape; everything was going well. Why should we destroy it with such dangerous, ugly-looking lines like that through our country. . . ."

The response of the power companies' lawyers was predictable. They tried to impeach the testimony of the witnesses, as, for example, when Ben Grosz had suggested that advertising was causing increased electrical demand.

MILLER: Have you caused any studies to be made or are you aware of any studies as to the dollars spent in advertising and promoting the use of electric energy as compared to dollars spent promoting the conservation of electric energy?

GROSZ: I can't comment on the exact figures, no.

MILLER: You have never seen any studies actually, have you, Mr. Grosz.

GROSZ: No. I have not.

MILLER: I have no further questions.

The power company lawyers also questioned the farmers about how they used electricity, suggesting they might not be practicing what they were preaching about conservation. For example, Jim Clowes, president of KTO, testified that he had been averaging fourteen hundred kilowatt-hours per month on his farm, but thought he could reduce it to perhaps a thousand kilowatt-hours. UPA's Miller evidently had access to the records of Clowes' account with Stearns County Electric:

> MILLER: You have given us some figures on your energy consumption for the last five years; you said it has been relatively even. You did experience a big jump in electric consumption betweeen '68 and '69, did you not?
>
> CLOWES: I wasn't aware that I had.

The response of the Energy Agency to the public testimony was revealing. For example, Richard Wallen's cross-examination of Jim Nelson suggested what the agency thought of the potential for conservation:

> WALLEN: Do you have any knowledge of any documents that go into a detailed analysis of why in 1974 we used less electricity than we did in '73. . . ?
>
> NELSON: No, I don't. I would guess that it is most likely due to the conservation information put out by the government and, possibly, some of the utilities.
>
> WALLEN: What type of conservation?
>
> NELSON: Well, I suppose that as for conservation, there is a slight cutback, that's probably—well, I really couldn't tell you what it is that people cut back on. . . .
>
> WALLEN: Would you say that turning down one's thermostat down to sixty-eight degrees from let's say seventy-four degrees was one of those?
>
> NELSON: Yes, I would say it's one of them.
>
> WALLEN: Furnace might run less?

NELSON: Yes, it certainly was a factor. I couldn't tell you how big a factor.

WALLEN: Would you call that factor a one-time effect?

NELSON: Oh. I see what you mean. I suppose it is a one-time effect. . . .

WALLEN: Are you knowledgeable in relationship to any studies relating GNP, gross national product, to use of energy?

NELSON: No, I am not aware of these studies if there are any.

WALLEN: No further questions.

NELSON: If I could just elaborate on one answer. I think we have only scratched the surface of what we can do in conservation as far as adding more insulation and finding more efficient ways of doing things. I think the turning down of the thermostat is in itself a one-time thing. But I think that's only the surface of what can be done.

Near the end of the day, Mark Walsh, a resident of Alexandria, took the stand and presented an alternative vision of the role the Energy Agency might play in these hearings. He began with a quote from a book, *Bread for the World*, by Arthur Simon:

Consider the example of energy. U.S. air conditioners alone consume as much energy as does the entire nation of China with its nine hundred million people. On a per capita basis, we use twice the energy that West Germany does and three times that of Japan . . . this country wastes as much as Japan uses. Turning down our thermostats and driving less may have an integrity of their own, but even if widely practiced, they are not [going] to cut back extensively on energy wastes. To be effective, energy conservation must result from carefully designed national policy, have the force of law, and be applied to the nation as a whole. We need then to move from the personal to the public realm on such a matter.

Walsh concluded, "I think the hearings such as this offer an opportunity to do just that. We have before us state people, people from the state energy office, and I think they can help to do just what this book says."

In all, fifteen citizens testified that day in Alexandria. At the conclusion of the hearing, Mullin thanked everyone for coming:

"I want to thank all of you for being here today, for offering your very helpful testimony . . . many things that you have brought to our attention. . . . You were a good audience. . . . I want to thank you all for that. I hope we have helped to educate you about the processes of the Minnesota Energy Agency and about this proposed powerline and about your state government. Thank you very much for coming."

The Conservation Alternative

The intervenors brought in only one expert witness. John Gostovich, a young agricultural economist on the staff of MPIRG, who had been serving as an intervenor, stepped around the table to testify on December 29. The focus of his testimony was conservation.

Gostovich began by quoting from the act that established the Energy Agency: "The legislature seeks to encourage thrift in the use of energy and to maximize use of energy efficient systems thereby reducing the growth of energy consumption, prudently conserving energy resources and assuring state-wide environmental protection consistent with an adequate, reliable supply of energy."

Thus, he asserted, the proper stance of the agency in reviewing applications for new energy facilities should be one of open skepticism: "It is the responsibility of the Energy Agency to grant need certificates only as a last resort, and only after a careful study of all alternatives, including energy conservation."

Gostovich criticized the CPA-UPA application on several counts. First, the demand forecasts might be self-serving; independent forecasts should be made. Second, the application contained no serious examination of the potential for conservation; the possibilities for reducing consumption through such means as pricing and end-use efficiency should be included. Third, the application contained no discussion of the potential for load management; the possibilities for reducing peak demand by offering incentives to consumers to shift to off-peak hours could substantially affect the need for new generating capacity.

He pointed out that the price of the CU project, projected to serve 260,000 accounts at a total cost of $700 million was $2750 per account. This "shows the amount of capital that could be invested at points of end-use to reduce energy usage. The investment of $2750 per account could insulate, retrofit existing systems, pay for peak load metering systems, pay for radio-controlled load shedding devices, install heat pumps, coordinate and enlarge energy conservation planning, and the like. In fact, careful study would show that the conservation pay-off of much smaller investments would yield ample reliable power for the CPA-UPA ultimate consumers."

The heart of Gostovich's position hinged on studies—ones that had not been performed. The power companies had not chosen to; the intervenors did not have the resources; the Energy Agency was the key.

The Agency Position

On January 13, the last day of the hearings, Dr. Richard Wallen, director of the certificate of need program, testified for the Energy Agency. He was the agency's expert on electric utility matters.

In Wallen's view, the situation was clear: "It has been demonstrated, I think quite adequately, that neither Cooperative Power nor United Power has the capacity to meet its current system load."

Wallen's prepared statement focused on the "near term," the next five years. In that period, he testified, the CPA-UPA demand forecasts were accurate, and if they erred, it was probably on the low side. The possible new developments he could see affecting demand in the near term were a tightening of natural gas and oil supplies and an increase of the price of natural gas, oil, and propane relative to the price of electricity. Both these developments would bring about electric demand greater than the historical projections presented by CPA-UPA.

During cross-examination, Wallen admitted that if electricity demand kept increasing at anything like the projected rate, it would soon outstrip generating capacity and the CU project, with its one thousand megawatts, would make little difference. Wallen predicted an electricity crisis in Minnesota within ten years: "There will be economic and social disruption statewide. . . . I foresee some sort of rationing of electricity in the next ten years."

The most striking difference between Wallen and the intervenors was his attitude toward conservation, as illustrated in this exchange with Duranske:

DURANSKE: With respect to forecast methodology, are you satisfied that UPA-CPA have considered in their application the impact of conservation? . . .
WALLEN: I don't know of any conservation methods, or conservation procedures, that they should consider that are applicable to their system at this time in the near term.

This was a curiously passive position in light of his dire prediction of a state-wide electricity crisis within ten years.

Wallen later submitted proposed findings of fact to Hearing Officer Mullin that recommended approval of the certificate of need application. That is what Mullin recommended to Energy Agency director John Millhone. On April 2, 1976, Millhone granted a certificate of need to CPA-UPA.

In his document granting the certificate of need, Millhone focused on the fact that CPA-UPA's anticipated demand would exceed their assured supply. The possibilities of purchasing power from other MAPP utilities and reducing demand through conservation were given short shrift in the document.

There was only one small paragraph dealing with purchase power:

Purchases of electricity from other sources is not available as an alternative to the proposed facility. No evidence was offered of the availability of such power.

Conservation was also dealt with very briefly:

Intervenors other than Agency staff suggested several factors such as price elasticity, changes in rate structure, end-use efficiencies, load management, and saturation levels of end-use appliances, which might make future electric demand decrease or grow at a lesser rate than the Applicants' historic trending indicates. No evidence was introduced that such factors would have a significant impact in the short term.

Existing state and federal conservation programs and any possible new programs are not likely to have a significant impact on the Applicants' energy and demand projections for the short term.

In retrospect, it seems that one possible alternative strategy purchasing power from the other MAPP utilities in the short term, say five years, and thereafter reducing demand through conservation and alternative energy sources—was never seriously explored in these hearings. It may not have turned out to be a viable alternative, but it certainly should have been closely examined. The fact that this and other alternatives were not seriously explored can be traced to some fundamental defects in the certificate of need process.

First was the mismatch of resources. The Energy Agency's position about alternatives to the power companies' proposal was that the burden of proof is on the intervenors. The problem is that invariably the intervenors do not have the resources to perform the necessary studies, to develop detailed analyses, and to map out substitute scenarios. Without such alternative studies, analyses, and scenarios, the energy companies have the field to themselves.

Second was the lack of a long-term plan for energy that explicitly included conservation and renewable sources. What the intervenors were able to do was to describe qualitatively a variety of mechanisms, such as changes in prices, insulating and otherwise tightening homes, load management, substitution of solar heating, wood stoves, or windmills, and so on, that might reduce total or peak electric consumption. With access to appropriate resources they might even have been able to quantify how much effect each measure could be expected to have. For example, there have been many studies of just how much energy consumption is reduced as prices rise or homes are insulated. But what would still be lacking would be an agreed-upon program of conservation and substitution, on the basis of which credible, quantitative forecasts could be made of just how much the anticipated demand is likely to be reduced.

A comprehensive, state-wide, long-range plan for reducing energy consumption would be required, with active programs of conservation and alternative energy. If such a plan were developed, with quantitative, year-by-year targets to judge progress, then it would be possible to integrate conservation and alternative energy sources into a forecast of anticipated demand:

anticipated demand = projected historical demand – conservation target – alternative energy target

Of course, in its early years, such a plan would undoubtedly have to be tuned; targets would have to be adjusted in light of experience. But over time, with continuing public input, a realistic state energy policy would evolve.

Both the task of redressing the mismatch of resources and the task of long-term planning that includes energy options extending beyond the interests of the established energy companies require the active intervention of the state. Leaving the information and the analyses and the planning in the hands of the energy companies is a prescription for little change in the way our nation does its energy business.

The Great Coal Mystery

To conclude this discussion of the need hearings, consider a mystery that the hearings might have helped to solve, but did not. One of the principal criteria to be used by the Energy Agency in deciding whether or not to issue a certificate of need was "possible alternatives to satisfying the energy demand." One alternative to the combination of a mine-mouth power plant in North Dakota and a powerline to Minnesota (the "North Dakota option") was a power plant in Minnesota, burning Montana or Wyoming coal shipped in by rail (the "Minnesota option").

The power companies claimed to have considered that alternative. They submitted for the record in the need hearings a detailed study performed by the Kansas City consulting firm, Burns and McDonnell, that compared the North Dakota and Minnesota options. The Burns and McDonnell study, completed in 1973, concluded that the North Dakota option would be less expensive than the Minnesota option and hence economically advantageous for CPA and UPA customers.

The key to the North Dakota option's economic advantage was the low cost of North Dakota lignite. At the time of the study in 1973, North American Coal had informed REA in a letter that it would charge 19.1 cents per million Btu for lignite coal at the Underwood, North Dakota, site. This was much cheaper than the

43.1 cents per million Btu that Montana coal delivered to Minnesota would cost.

For the North Dakota option to be economically attractive, the North Dakota lignite *had* to be much cheaper. According to the Burns and McDonnell study, the capital cost of building a power plant in North Dakota and a powerline to Minnesota (estimated to be $536.7 million) was much greater than the cost of building a power plant in Minnesota (estimated to be $306.1 million):

> The Minnesota load center location has a substantial capital cost advantage over the North Dakota site. This is primarily due to the large transmission cost required to deliver energy from North Dakota to Minnesota, a substantial increase in boiler cost required by the lower heat content lignite fuel and a larger net plant capacity in North Dakota—approximately 11% greater—required to compensate for these differences.
> . . .
> With this large capital cost difference, it is evident that there must be a substantial saving in fuel costs at the North Dakota mine-mouth site over the rail-delivered fuel cost at the Minnesota load center site to make these alternates equal in total power costs for the same quantities in delivered energy.

If North Dakota lignite cost 19.1 cents per million Btu and rail-delivered Montana coal cost 43.1 cents per million Btu, the operating cost advantage of the North Dakota option would more than compensate for its initial capital cost disadvantage: the Burns and McDonnell study estimated that the total cost of supplying power for ten years with the North Dakota option would be $618.2 million as compared to $691.4 million for the Minnesota option. On these grounds the North Dakota option was recommended.

After the Burns and McDonnell study was completed, but before the CU project was underway, however, North American Coal Company changed its terms dramatically. It told the REA that it would charge $1.00 per million Btu for its lignite, not 19.1 cents, and furthermore, it demanded that the power companies finance the coal mine, an unprecedented move that would cost an additional $200 million.

Frank Bennett, director of REA's North Central operations, blames North American's sudden change on the Arab oil embargo of November 1973: "That just turned the coal world topsy-turvy. I have no sympathy for the coal companies. I think they . . . used the Arab oil embargo for their own selfish interests to increase the cost of coal. . . . Well, believe me, it made them much more difficult to deal with and as a result they rescinded their other letter. . . . They felt they held the trump cards. Probably did. They said we'll have to finance thé mine for them. . . . It wasn't our desire. It wasn't the way the project started."

It is clear from the Burns and McDonnell analysis that if these terms were accepted, the North Dakota option would cost far more than the option of building the power plant in Minnesota. Not only would the capital costs be much greater, the operating costs would be higher as well. Nevertheless, CPA, UPA, and the REA office in Washington did choose to accept these terms in 1974. CPA and UPA signed a coal agreement with North American's wholly-owned subsidiary, Falkirk Mining Company, specially created for this purpose. Why the power companies and the REA chose to do so is perhaps the major unsolved mystery of the CU powerline story.

Had they known, the co-op members, who in principle decide such matters in a democratic cooperative, would certainly not have made the same choice. As the customers of CPA and UPA power, however, they would ultimately pay.

All of this had occurred by the time the Energy Agency came to consider the question of whether the CU powerline was needed. As noted, the Burns and McDonnell study was submitted as a hearing exhibit by the power companies. However, the intervenors were evidently unaware of the contents of the coal agreement and the adverse economic impact on CPA and UPA customers of the decision to proceed with the North Dakota option once North American Coal had changed its terms. They did not press the issue of this most puzzling behavior by the power companies and the REA. What Verlyn Marth frequently referred to as "a scandal bigger than the Lockheed loans" was never seriously reviewed in the need hearings.

The Route Selection Process

While the need hearings were going on, the process to choose a route was beginning. On December 9, 1975, the EQC accepted CPA-UPA's application for a construction permit. It had six months to choose a one-kilometer-wide route within the corridor. The mechanics of route selection were the same as for the corridor—a citizens' committee, information meetings, and public hearings, with the EQC making the final decision.

As far as the farmers were concerned, the formal process of route selection was simply the final nail in the coffin. The question to be addressed was not whether the line would be built, but exactly where it would be built, that is, what farmers would get the line.

That was the formal purpose. At the same time, it did give the farmers one more opportunity to express the depth of their opposition and six months more to organize that opposition.

Many of the members of the route evaluation committee were farmers from within the corridor. Included were several of the protest leaders, Kathleen Anderson and Noel Kjesboe from Grant County, Harold Hagen from Pope County, and Virgil Fuchs and Mary Jo Paulson from Stearns County; Michael Howe was also named.

The EQC had learned a lesson from the corridor committee; it designated its own chairman, Harold Josephs, a farmer and teacher, with a background in electrical engineering, before the first route committee meeting.

The protestors had learned some lessons, too. At the first meeting of the route committee, Harold Hagen told his fellow committee members how much influence they might expect to have: "We were told at one time that the citizen advisory committee ranks very high in their input. . . . The results of our corridor selection committee lead me to believe that we rank high only if we happen to decide the same as the hearing officer and the EQC decide. That's how we rank high. . . . We don't rank high and our batting average won't be anything unless we happen to decide the way they want us to decide."

The farmers on the committee had come a long way from their naive hopes at the outset of the corridor process; Charles Carson's advice to "challenge the EQC itself, hold a press conference and denounce its pandering to the power companies" no longer seemed quite so radical. Soon after the first meeting, they sent out a press release, headlined ADVISORY COMMITTEE MEMBERS SAY THAT THE PUBLIC IS BEING BILKED, to eighteen newspapers from one end of the corridor to the other. The release stated:

> The public is being hoodwinked into thinking that trans-mission lines are routed with maximum public input. To the contrary, the committee members stated that the MEQC staff and the power utilities themselves are calling all the shots. . . . It was felt by the committee members issuing this release that the MEQC will give the power utilities exactly what they wanted in the first place and that the State Agency is merely being used as a "front" for the power industry.

Others on the committee were disillusioned, too. For example, after the first meeting, Michael Howe received a letter from another member, who had not previously been active in the protest:

> I am the fellow who talked briefly with you after the citizens advisory committee meeting in Alex a couple of weeks ago. At that time I mentioned the idea of getting as many mem-bers as possible to stage a "mass resignation" and expose, in the press, the committee's "rubber stamp" status. I later talked with another fellow who had overheard a group of people talking like they were ready to resign now (after the first meeting, that is). So it looks like there are more than a few people who are unwilling to play the EQC's game.
>
> I know you were thinking about getting the committee to propose as many routes as possible in order to get more people in the corridor upset and involved. I still think that's a good idea, especially since the EQC won't accept the proposals of the committee anyway. Simultaneously, however, I am wondering if it would not be advantageous for those of us opposed to this whole affair to dissociate ourselves complete-ly from the routing procedure. I am sure that the committee, or what's left of it should a group resign, will still propose routes. But by that time we would not be a part of it, and would have simultaneously publicly voiced our objections to this "game."

What the committee members could do was very limited. As the press release put it:

The present committee . . . has been given almost no opportunity to propose a route within the designated corridor. While the power utility and the EQC staff have been studying data for months to develop a route through the designated corridor, the Routing Committee was told on December 16 that all of the routes to be evaluated shall be defined by January 23 and that the Routing Committee is encouraged only to evaluate routes proposed by the EQC staff or Coop-United.

On January 23, the possible choices for the route were set by the EQC. Eighty "route segments," interconnecting links within the corridor, were designated. The actual route would have to be some combination of these route segments.

The other phase of "public participation" began with two weeks of information meetings in late February. Jim Nelson remembered the meeting at Elbow Lake:

The information meeting began by the EQC staff members introducing themselves and describing their position briefly, and at that point the information meeting was turned over to the applicants who proceeded to show two films. The first was a slide film with a Mr. Don Jacobson reading from the manuscript and the second film was a slide tape. Both of these films while possibly containing some factual matter were in the main propaganda devices. The second . . . was a film accompanied by music which was of a generally happy, cheerful nature and showed the people with smiling faces and generally was an attempt to make you feel kind of good all over as far as I could tell. Then after this was accomplished a commentator began to say very nice complimentary things about UPA-CPA and their project. This I would say was the main, eighty to ninety percent, of the information meeting. After this there was a period for questions, some of which were answered by the EQC staff members. Most were responded to, not entirely answered I would say, by the Applicants. . . .

Then came three weeks of public hearings, one evening in each county within the corridor. They were conducted by hearing officer Ted Mellby, an attorney from southern Minnesota.

The first public hearing was held in Anoka, north of the Twin Cities, on March 23. First to testify were experts from Commonwealth Associates. It soon became clear that the purpose of these hearings was to iron out the details. Witness this polite exchange among the EQC's Emmett Moore, UPA's Roger Miller, and Hearing Officer Mellby during testimony by Commonwealth's Robert Folmsbee:

MOORE: On page five of Mr. Folmsbee's statement, the second complete paragraph which reads as follows: "Cleanup of the right-of-way work areas will be continuous throughout the construction period; after all the work has been performed, the final cleanup operation will proceed down the right-of-way to restore it to substantially its former condition. All waste and scrap material will be removed, ruts and holes graded. All areas where natural vegetation has been removed will be reseeded with a native grass as soon as practical after construction." My question is . . . this, are the applicants willing to have statements, these statements, put into the construction permit?

MILLER: Dr. Moore, we would be pleased to have statements, these statements, put in the construction permit.

MOORE: Thank you.

MELLBY: Dr. Moore, the language that was agreed upon in the conclusion of the construction permit, was that limited to the first sentence? Paragraph three? Page five?

MOORE: No, the entire paragraph.

MELLBY: Thank you.

The most important testimony on that first day of the hearings came from Commonwealth's Robert Cupit. With eighty route segments to choose among, on the surface at least it appeared that there were a large number of possible combinations that could be chosen to make up the route. However, this was largely an illusion.

Cupit's team at Commonwealth Associates had performed a detailed analysis of each of the eighty route segments, compiling information on fifty-five features that related to EQC exclusion, avoidance, and selection criteria. The fifty-five "inventory features" determined for each segment ranged from "residences, number in route" to "Minnesota DNR wildlife areas: length of centerline crossing (feet)" to "federal waterfowl production areas: acres in route," to "potential irrigable areas, length of centerline crossing (miles)."

Based on his analysis, in light of the EQC criteria, Cupit announced that eight of the route segments were clearly not acceptable. These he labeled "high constraint" segments. Five other "connecting links" between these segments could also be ruled out. Thirteen segments eliminated out of eighty may not seem like a large number, but the net effect was to limit severely the possible options. There were really very few possible deviations from the CPA-UPA preferred route.

The way the remaining hearings were arranged, the farmers could fight among themselves over who would get the line. A typical hearing would begin with Hearing Officer Mellby calling on Cupit to describe the route segments in that county. Then Mellby would call out the segments one by one and individual farmers whose land lay on each segment were invited to come forward and say why they should not have it. Jim Nelson was one who understood what was going on: "the theory of the route hearing was to—as I see it—was to hopefully get farmers played off against one another. They would have different routes drawn out and hopefully the farmers would get up and say it shouldn't be on my land but it should go on this other route over here for whatever reason he could dream up."

In Grant County, the farmers refused to play the game. Nelson described the hearing at Elbow Lake: "At the beginning of the hearing, the hearing officer . . . said, 'Is there any testimony on this route segment?,' and then people who lived on that segment were supposed to get up. There was no testimony on any individual segment and finally he got to the last one and he acted as if he were going to close down the hearing even though there was a big crowd there. So at that point I said I had some testimony that applied to all the routes. Everybody followed and the farmers talked together on that—they didn't fall into this trap of trying to push the line one way or another."

What happened at Elbow Lake was a demonstration of opposition. Hearing Officer Mellby was told by two witnesses that the line should be routed at the bottom of the Minnesota River. He was also challenged by Kathleen Anderson about health and safety, a topic the EQC had ruled out of bounds for the route process. His response really stirred up the crowd:

> MELLBY: . . . If you are going to prove that this line is not safe that will be your burden.
> ANDERSON: Isn't it up to the power companies, since they are building the line, to prove that it is safe?
> A VOICE: Yes.
> MELLBY: No.
> SEVERAL VOICES: Yes.
> MELLBY: I appreciate the difference of opinion.

Grant County was an exception, however; in most places the farmers played the routing game by the EQC rules. Only in Pope County did it make any difference. There a choice was made that would have a profound impact on the course of the protest.

The solid line in figure 5 shows the preferred route through Pope County and western Stearns County that CPA-UPA submitted to the EQC. That represented a slight northward shift from the route that had been informally negotiated with the Bonanza Valley Irrigators in 1974 during the Pope County planning commission hearings (dotted line).

At the time of the Pope County route hearing, in Glenwood on April 7, 1976, the prevalent belief among local farmers was that the power companies would get the route they wanted. What was not known, however, was that (even if Cupit's high constraint segments were eliminated) there was another feasible route through Pope County to the north (dashed line).

At the hearing, opponents along the preferred route turned out in strength. They had been organized since the Pope County planning commission hearings and had been lobbying vigorously against the line down in St. Paul ever since. They were prepared with a barrage of testimony. The Glenwood city council told of favorable reaction from the state in response to initial feelers about possible relocation and expansion of their airport. The Bonanza Valley Irrigators explained the anticipated expansion of irrigation throughout the valley and the devastating impact the powerline would have. The supervisor of the Glacial Lakes State Park described the negative visual effect on the large numbers of campers who used the park. As their segments were called, many farmers along the preferred route took the stand with impassioned statements against having the line come across their land.

In contrast, that evening in Glenwood there was hardly a stir from the unsuspecting folks up by Lowry on the northern route. A few came forward to make inquiries and express concern, but they were not organized and there was no sense of urgency.

Meanwhile the route committee was hard at work. It met frequently, hearing reports from individual members, who inspected every one of the eighty route segments. When it was time to make a recommendation, it became clear that the only major choice was between the northern and southern

Figure 5. The solid line is the route UPA-CPA told the EQC that they preferred in Pope County; the dotted line is the route they had requested before the Pope County planning commission; the dashed line is the Pope County route the EQC selected.

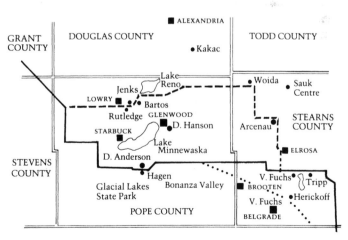

routes in Pope County. When a vote was taken, the committee
was evenly divided. The committee report explained the reason
for the split in objective terms:

> The committee ranks as having equal impact a route consist-
> ing of [segments] 24, 30, 34, 35, 37, 46, 49 and 51 with one
> consisting of 25, 27, 29, 39, 40, and 42. Both of these combina-
> tions have visual impact on recreational lakes, both are pre-
> dominantly east-west crossings with significant length of
> diagonal crossings, both cross considerable amounts of ir-
> rigable land, with some active irrigation. Both would require
> approximately the same number of deviations from the
> centerline to avoid homes and lakes. The most significant of
> these in the northern route are near Belgum Lake and Erick-
> son Lake and along the Little Chippewa River in RS 24 and
> near Leven Lake in RS 37. In the southern route, the most
> significant deviations are required in the area south and west
> of Sedan in RS 39. The northern route is the longer by 3.89
> miles. The southern route has significant impact on Glacial
> Lakes State Park.

Other, less objective, factors may also have been involved
in the vote. Many of the committee members had an interest in
the choice. According to Michael Howe, "a great majority were

Figure 6. The CU powerline route in Minnesota.

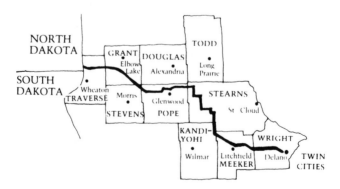

pushing for a route as far away from home as possible." Virgil Fuchs echoed his words: "of course everybody was voting to keep it as far away from his home as possible."

On June 3, 1976, the EQC issued a construction permit to CPA-UPA and designated a route (figure 6). In its "Findings of Fact and Conclusions," the EQC presented an analysis that precisely paralleled that presented by Commonwealth's Cupit at the first public hearing, right down to the last "high constraint" segment and unacceptable "connecting link." The route designated was essentially the CPA-UPA preferred route, with one notable exception —in Pope County, where the northern option was chosen.

Suddenly the line crossed forty-nine new miles of farmland. It passed by some new towns like Lowry and Sauk Centre and Elrosa. It got put on some new farmers like Dennis Rutledge, Scott Jenks, Tony Bartos, Math Woida, and John Tripp. By the time they learned what had happened, there was nothing they could do about it. The public had participated; the state had decided. It was all over but the actual construction of the powerline. Or at least that is the way it was supposed to be.

Jim Nelson

Profile

Jim Nelson and his wife Cheryl live on an 800-acre farm in Grant County, five miles south of the town of Elbow Lake. A long driveway leads into a beautiful wooded grove, which surrounds their home and that of Jim's parents. All that mars an idyllic setting is a succession of enormous powerline towers that march diagonally across their field.

As we drove into the farm, we were struck by how many towers are on your land. In retrospect, how do you feel about it?

You mean about the line?

Yes, once its there and you can actually see it.

I suppose it varies with people who've been fighting the line, but every time I see the towers, every time I walk in the fields, it kind of brings back all the memories of fighting the thing, and in a way every time I see it, I feel more bitter. The power companies—I just kind of assumed they were the enemy—but what disturbed me the most was the state was supposed to be independent. I was never too naive about what the state's role would be, except I never thought it would be as bad as they were. I didn't expect them to be completely independent, but in the end, I thought they would give a little more than lip service to the protesters and they never did. The whole story of the line, the way it seemed to me, was that they would pretend to do something. Anyway, whenever I see a tower or work around it in the field, it brings it up all over again.

What lessons have you learned about what to do or not to do? About what works or doesn't work?

Having had a while to think about it, I think where I was probably misled, well, not misled but misdirected, I always had the
feeling that if we'd won it, we'd have to win or lose in the courts.
Now I think you have to work in the courts to make the whole
thing legitimate—otherwise people will say, "What are you doing
out here in the fields? You haven't tried all the channels within
the system. You know, you're not even trying to work in the system, so what are you doing out in the fields?" So you kind of do
have to work in the courts I think, but I think where I was wrong
was I probably put too much emphasis on that and I think the real
emphasis should be out in the fields. That's in fact where we won
or lost—in the fields, because there's nobody really even took us
serious until people were out in the fields. So I guess if I've learned
anything it would be to sort of just work in the courts and not to
appear to be illegitimate, but put most of the emphasis out in the
fields. It would be a lot cheaper, too!

*In April of '75 the utilities gave up their grandfather privilege
and the Environmental Quality Council and Energy Agency assumed jurisdiction and held hearings. Can you tell us about
those hearings?*

Right. Some people will tell you, and I suppose it's different at
different places along the line, that at the initial hearings the attendance wasn't too great. In this area—I know in Pope County
and Grant County, and in Stearns County, there was very large
attendance and there was a lot of participation. The meetings
would generally go past midnight and people, given that they did
participate to that extent, it was probably implied that some of
what they said would probably be heard. In the end, of course,
people felt that none of what they said was really heard.

You think the hearings amounted to mere formalities?

Right. Just as example of this, John Millhone [director of the
Energy Agency] stated at a press conference in Fargo, North Dakota, in October of '76 that they never did consider the alternatives to the line, because the power company had already started
building the plant in North Dakota. And that's basically the game
that the state agencies played the whole way. They never seriously considered whether the line was needed or serious alternatives

to line routing or line design or anything. The power companies said, "Well, we've already bought the parts and all the pieces for this." So in the end the state officials were afraid to—I don't know if they were afraid of the legal might of the power companies, but they felt the power companies had already started doing this and who are we to ask them to do something different from what they expect to do?

Do you think the farmers should have handled the hearings differently?

At the corridor hearings we had counsel, but on the route hearings we didn't. We didn't have legal counsel at the route hearings because we couldn't afford it and because it didn't seem to make any difference at the corridor hearings. I guess to me in retrospect, I think the farmers handled the hearings probably the best way possible. I think that in particular on the route hearings, when I think back to one they had in Grant County, the theory of the route hearings was to—as I see it—was to hopefully get farmers played off against one another. . . . I think the farmers handled the hearings probably as well as they could. The fact is that there were never any open hearings—that the decision had been made before the hearings began.

Do you think it would have helped if the farmers had protested at the hearings since you think the hearings were a foregone conclusion?

Actually what they were doing at the hearings was protesting. It was a nonviolent protest. At the corridor hearings in Elbow Lake we had big signs outside, and people marched in picket lines. The hearing examiner then asked people to put their picket signs away. He said he thought it was a bad deal to have them inside the hall—it destroyed the objectivity. And people obeyed him—they put them down on the floor. The testimony at the hearings, of course, was often emotional as well as factual, and the intention, by the size of the crowds at the hearings, was to show the state how strong resistance was. And there part of the problem is that the EQC officials didn't show up at these hearings. They weren't required by law to be there—they left it all to the hearing examiner. So the people in the state capitol—they never really got out-state much. I think Peter Vanderpoel went to one hearing. They never believed things were as serious as they actually were out here because they never came out and took a look.

Would you mind giving us a little of your background?

I got a bachelor's degree in physics and a master's degree in nuclear physics. I grew up around here. I wasn't born on this farm, but I was raised on this farm and lived here all my life except for when I went to the University of Minnesota and at the University of Pittsburgh in nuclear physics. Then I worked for the Raytheon Company for three-and-a-half years, so all told I suppose I was gone for eight-and-a-half years. Then in 1972 Cheryl and I came back to the farm.

Why did you decide to come back?

Just missing, for one thing, being able to look out a window and seeing for a mile and to live under less crowded living conditions. I never really thought about living in a city; I just assumed that some day I'd work as an engineer and live in a city. If I'd really thought it through in detail, earlier in my life, I probably would have realized that I didn't really care for living in a city right next to a lot of other people. Then the other thing was that engineering, at the time that I quit, was really not going anywhere. Government funding had been cut way back, all that we were doing at Raytheon really was upgrading old radar systems. There wasn't any new research and development; it was purely taking an existing system and trying to make it a little bit better. So it wasn't very interesting.

The decision to come back to farming—are you glad now?

Oh, yes. The powerline's the only bad part about it. Other than that, I've been happy with that decision. I think it was a good one. I like the general life of a farmer, the way a farmer lives and works.

And now you farm with your dad, too?

Yes, we farm together eight hundred acres. Although he's technically retired, he still works. We plant corn, soybeans, wheat, and barley, and we also raise at times flax and oats.

Do you have brothers and sisters up here?

Yes, my brother teaches at the University of Minnesota. And

then I've got a sister who is a doctor. My oldest sister married a rancher in Montana and she teaches high school English.

What would have happened to the farm if you hadn't come back?

Well, I'm not absolutely certain. I guess what would have happened is that my dad would have rented out the land, but after that I don't really know. That is part of the feeling of coming back; you sort of want to keep the farm intact and in the family.

How would you describe yourself politically?

Well, it's kind of hard because there are so many different issues. On local government I guess I am a conservative—that as much as possible should be done by local governments. I guess that is a conservative position, but as far as social welfare, taking care of people that can't take care of themselves, I would say that I'm a liberal. It's very hard to put labels, but I know that my feelings on local government would be classified as conservative because historically that is a conservative position.

How has this powerline struggle affected you? What changes have you felt?

In a way that is kind of a hard question, because I suppose a person never took stock of himself four years ago, so I don't really know what I'm comparing it to. But I suppose that I've always been a skeptic of government, always felt there's more to government than what you read in the fifth grade history book. I would have to say that my skepticism about government has increased a lot more; it's a lot greater than it was four years ago, that the decisions are not necessarily made for the right reasons. There's a lot more that goes into those decisions than should go into them, more than I suspected four years ago. I didn't feel four years ago that I was a babe in the woods, but this struggle has given me a greater feeling that things can't be allowed to go along and that you can't leave some things up to someone else. You can't really trust a state agency or leave a decision up to some entity and say, well, they've got the expertise and the knowledge and hopefully the conscience to deal with the issue and I'll let them come up with the right answer. You can't do that, you've got to be watching government, state agencies, everything that affects your life.

How do you feel about the future for farmers? Some of the people we've talked to have been pessimistic. Do you share this pessimism?

I share some of it. I'm concerned with this idea of running corridors of powerlines across farmland and just the general encroachment on farmers and the land. If there's uncontrolled development, it doesn't look too bright from that standpoint. But on the other hand, I'm hoping that with enough movements like the powerline movement, that maybe we'll get into alternative sources of energy a little quicker and so one could be a little more optimistic. I guess I'm basically optimistic about it. A lot of people besides myself have said it, that to be a farmer, you have to be optimistic. You put the seed in the ground, you're optimistic you're gonna get rain and it's not gonna get hailed on. If you're pessimistic, you wouldn't last very long.

The power co-ops think they have a right

To take a farmer's land without a fight—

We don't want no powerlines around here,

We don't want no powerlines around here,

We don't want no powerlines,

We don't want no powerlines,

We don't want no powerlines around here!

—from "We Don't Want No Powerlines
Around Here" by Russell Packard

4   On the Brink of Violence

The farmers were unsure about how to respond to the EQC decision. One obvious avenue was the courts. CURE's lawyer, Norton Hatlie, vowed he would appeal the decision in district court and seek an injunction halting construction of the line. Another possibility was action by the state legislature.

As they met to discuss what could be done, opposing the line in the courts or in the legislature were ideas that came naturally to many of the leaders, who were convinced these were the correct ways to seek change. Jim Nelson has reflected on how the leadership viewed their options. "I think that in fact the court strategy is the only real alternative as long as you make a decision that you want to work within the system, as opposed to working outside the system. At the time when we began, the people who were involved in the organizations were upstanding citizens in the community, chairmen of hospital boards, township boards, and that kind of thing. They were the respected and responsible members of the community and they were not people who have any easy time or even have any possibility of resorting to violence or disruption, so that really the court strategy and working through the legislature were the only real options, and the legislature was too slow; we knew that, so the courts were our main hope."

Some of the farmers, however, were no longer in a mood to work within the system. Many had expended an immense amount of time and energy "participating" in the State proceedings, which they now felt to have been a charade. Virgil Fuchs, who had attended nearly all the information meetings and hearings and who had served on both the corridor and route evaluation committees, articulates the anger and bitterness that many farmers felt: "To put a powerline on you is bad enough, but to go and make you go through all this torture beforehand and then give you the son-of-a-gun, that's something else. . . . They say they're giving the people input or whatever; they'd be a lot better off just to decide to put it over here, because we had no say-so anyway."

Tuesday, June 8, 1976, was a turning point in the powerline struggle. That was the day Virgil Fuchs took matters into his own hands.

As the surveyors were working their way across Fuchs' land early that morning, he drove his tractor toward them and smashed a tripod. Then he rammed the tractor into one of the company pickup trucks. In retrospect, he says, "Don't ask me why I did it. I wouldn't know. I don't know why I did it. I suppose a guy would think it was to bring to the public's eye what was going on out here."

The power companies filed formal charges against Fuchs and a Stearns County deputy sheriff drove up to his farm that evening. Fuchs was out working in his fields after dinner when the deputy arrived. He left word with Virgil's wife, Jane, that Virgil was supposed to come to St. Cloud the following morning. Virgil recalls, "At that time I didn't know what the hell was going on and he just told her I should be in St. Cloud. I didn't know where in the hell to go in St. Cloud, to Gamble's or Dayton's [two retail stores] or where. So I called the number he left and I said, 'This guy was here today and he told me to be here tomorrow,' and I asked 'Where is here?' And they were real snide. They said, 'You be here at nine o'clock,' and hung up the phone. And then I called three more times and I finally got through—'If you want me to come there, where am I supposed to come?' "

The next morning Fuchs drove to the county courthouse in St. Cloud. When he arrived, there was a big crowd outside. Word about Virgil Fuchs had gotten around and farmers from all over western Minnesota were in St. Cloud that morning, some having traveled over a hundred miles. Fuchs notes in retrospect, "People weren't used to arrests in them days you know."

The morning dragged on and Fuchs was not released from jail. It was a very emotional incident for the farmers. As they waited outside, they were incensed that "a good farmer was being treated like a crook."

In the courthouse, Fuchs was being charged with two felony counts, each carrying a maximum penalty of five years in jail and a five thousand dollar fine. According to Fuchs, "I think they thought, this will take care of the whole business. This is what they thought. They were really going to scare the hell out of the farmers and that is going to take care of it. It didn't work."

He refused to pay bail: "I asked them, 'What do you guys want? I'm not going to pay you guys no bail or bond or anything. If that's what you want, put me in jail. . . .' I just knew the farmers wouldn't accept it." Stearns County officials evidently recognized that the crowd outside would be very upset if he were jailed. After several hours, Fuchs was released without bail on the condition he not interfere further with surveying.

It was an important event. Without anyone's planning it, Fuchs' arrest united the protesters and pointed the protest in a new direction. This was the first time many of the farmers from Grant, Pope, Meeker, Traverse, and Stearns counties had ever met each other. Virgil Fuchs had stood up to the power companies. Farmers from all over western Minnesota had backed him up. They left St. Cloud that day unified and determined.

There is no clear evidence that Fuchs intended his action as an example for the other farmers. But his arrest was an inspiration to them. It came at a time when the farmers had been uncertain about how to carry their struggle forward. Alice Tripp, Elrosa farmer and neighbor of Fuchs, recalls the importance of his act; she believes it opened a new phase of the protest: "Virgil is quite a hero around here. Almost everything started with Virgil's running over the equipment. From then, people's involvement in our area became more intense."

When the surveyors came out the next week, the farmers responded in a way they never had before. Notified by CB radio, fifty to sixty farmers turned out en masse within a matter of minutes to block the surveyors' work. There was no violence; when the surveyors saw the farmers coming, they immediately left.

The Stearns County sheriff refused to give the surveyors protection until he received assurance the construction permit was legal. The farmers contended that the surveyors were violating a state requirement to wait sixty days after EQC approval to begin construction. Without protection, the surveyors were not anxious to confront the farmers, especially after the Fuchs incident. The companies decided to call off the surveying.

John Tripp remembers this first week well: "They [the surveyors] were at junction of highways 14 and 71 [near Fuchs' farm]. We never knew for sure why they started here. Some people thought it was because they had to have that particular point fixed for some reason or other to complete their survey. Others thought they made up their mind to break the resistance right at the point of resistance. It was Virgil and the younger guys who had their CB's and chased them. Some farmers would follow the surveyors and some would stay. We were the stay-there type. It was kind of scary. I guess you got the feeling you might be humiliated; you know, you're doing something wrong. So we stayed, and I remember one reporter kept asking and asking us, 'Well, what are you going to do? How are you going to stop them? What are your tactics going to be?' I couldn't tell them. We really didn't have a plan; we didn't know what we were going to do—we just knew we were gonna be there and not let them survey."

Stearns County sheriff James Ellering met with the farmers at Bucky's Bar in Elrosa on June 14 and tried to calm tempers. Al Bertram, an Elrosa farmer, told Ellering, "The thing that bothers me is living conditions—who the hell wants to live by one of these power lines?" Ellering asked the farmers, "Have you folks either collectively or individually contacted your local representatives?" Anne Fuchs, Virgil's mother, responded, 'We'll look into it, but they don't do nothing. We're just helpless; we're just helpless. What can we do?"

The farmers asked Ellering "where the law stood" in this conflict; "Does it side with the utilities or help the people?" He answered, "I wish I knew," and asked the farmers to understand his position: "Please understand right now; whatever the legal opinion is, I have to see it is acted out. This puts me in one hell of a spot!" The sheriff did promise to stay in close touch and requested John Tripp to act as liaison with his office.

On June 24, the surveyors returned, armed with an opinion from Peter Vanderpoel, chairman of the EQC, that the surveying in Stearns County was legal. The farmers were not persuaded. About seventy of them chased the surveyors away. A deputy sheriff warned the angry farmers that the surveyors would return the next day, with a sheriff's department escort if necessary. They replied that they would chase the surveyors off their land, escort or no escort.

Al Bertram confronted the surveyors in their car and warned them, "Take off or I'll take you apart piece by piece." He then turned to the deputy and said, "I've got teenagers that can shoot as straight as I can. . . . Let 'em come. You'd better build a bigger jail."

The surveyors did not return to work the next day. Instead, company officials met with Sheriff Ellering in St. Cloud to demand law enforcement protection. It was a dramatic meeting. Ellering said, "I don't want Stearns County turned into a battle ground. I ask you, how can we resolve this?" Phil Martin, UPA general manager, responded, "Only by not building the line."

Ellering maintained that, "Surveying is only going to propagate more problems," and added, "I would be deeply disturbed if we take decisive action on this and somebody gets killed." CPA attorney John Drawz retorted, "We're convinced if surveying were delayed by a week or a month, we'd be in the same condition. We have millions of dollars in this line and the project has to go forward. . . . Five hundred people are saying, 'Stop the world, march to my drum.' Do we allow them to take the law into their hands?" Jerry Kingrey asked Ellering, "Will you enforce the law?" Ellering replied, "With what we know would have happened out here today, how can I?"

The next morning, Saturday, June 26, Ellering sent a telegram to governor Wendell Anderson requesting him to intercede "by any means available," in the controversy. The governor, through an aide, expressed his concern, but indicated that for the time being there was nothing he could do.

On Saturday afternoon Sheriff Ellering again met with the farmers, at John and Alice Tripp's farm, near Elrosa. Alice Tripp recalls, "The wind was blowing and he stood on the back of a kitchen chair and he was up high and hardly anyone could hear him and there were horses and it was a real country-western, you know." The hundred farmers there wanted to know which side the sheriff was on. He first said he would have to enforce the law and that it was his job to protect the surveyors. The farmers argued it was his job to preserve peace in the community and the best way to preserve peace was to keep the surveyors out. When Ellering left, he said he would decide whether to provide security for the surveyors by the following Tuesday.

On Tuesday, the county courthouse in St. Cloud was crowded with power company officials, protesting farmers, and the news media, all waiting anxiously for the sheriff's decision. At 9:30 A.M. Ellering walked into the room. He refused to answer any questions and instead read a one-page statement. His position had changed. Drawing cheers from the farmers, Ellering suggested the EQC reopen examination of the line. He argued the problem was created when the construction permit was granted and that was where it should be resolved. He went on to say, "It [a resolution] cannot and will not be accomplished by a confrontation of the Stearns County sheriff's department. I will not point a gun at either the farmer or the surveyor."

The power companies were naturally very unhappy. Don Jacobson, UPA spokesman, announced the utilities would seek state law enforcement assistance if the county refused or was unable to provide it. However, the governor's office, not anxious to involve itself in this escalating conflict, indicated such a request would have to come from the county before the governor could take any action. At this point, Ellering would not make that request.

A stalemate developed. The Reverend George Speltz, bishop of the St.Cloud Catholic diocese, sponsored two all-day meetings between company officials and farmers, but they produced no agreement. In mid-July, CURE appealed the EQC construction permit in Stearns County district court. At the same time, CPA and UPA, admitting poor public relations, began an extensive newspaper and broadcasting advertising campaign to "re-educate" farmers on the need for the line. There were no further confrontations since the companies, without the sheriff's assistance, were unable to survey.

Work on the line came to a complete halt on August 11, when Stearns County district judge Charles Kennedy issued a stay prohibiting surveying and other work on the line in the county. Substantial portions of the testimony at two of the EQC route hearings were mysteriously missing from the official transcripts. The transcripts of the hearings at Paynesville and Litchfield seemed to have been selectively edited, with farmers' allegations of irregular practices on the part of power company easement agents removed.

The EQC could offer no explanation. EQC chairman Vanderpoel has reflected, ". . . it was clear that there was stuff missing. Now it isn't just that it was cut out at some point, because if you read them—and I had read them—they read fine. You didn't realize there was anything missing. And when they brought it up, . . . I remembered testimony at Paynesville which wasn't in there. . . . I don't know what happened—she [the court reporter] fell asleep or what the hell."

CURE's lawyer, Norton Hatlie, argued that since the record was not complete the construction permit should be declared null and void. Judge Kennedy ruled the EQC would have to make the record complete before any more surveying would be permitted in Stearns County.

There was a lull in the protest while the power companies, forced to leave Stearns County, decided where to survey next. They chose the junction of state highway 4 and county road 3 in Meeker County, directly south of Stearns County, and began work on September 28. The farmers were waiting for them there.

When the surveyors first arrived at 8:00 A.M., they were greeted by about thirty farmers. They watched as the surveyors got out of the truck, opened the back, and pulled out a tripod. Immediately they surrounded the truck and the crew. The crew radioed back to UPA headquarters for help. General manager Phil Martin and other UPA officials drove to the scene. Martin told the crowd that the surveying would continue: "I know you may not like the line and I realize you don't want it built, but we have gone through almost two years of state hearings on the matter; there were scores of information meetings and several court cases and now the matter has been decided. The state says we have the right to build the line and our one million members need electricity it will bring to Minnesota. It must be built." The farmers voiced their disagreement. "Those hearings were fixed," hollered one voice from the back of the crowd.

By early afternoon there were several hundred protesters on the scene. Martin asked Meeker County sheriff John Rogers to move some farmers who were sitting on the back of the survey truck. Rogers refused, saying "I'm only here to protect the farmers and the surveyors in case a fight breaks out." Martin was incensed. "I'm here to protect both sides," said Rogers. Martin then met privately with Rogers and Meeker County attorney Thomas Nagel. Afterward, Nagel was asked what had happened. "As I understand it, the power company intends to build a line through here and the farmers intend to prevent it," he replied.

There was no surveying that day. Late in the afternoon the surveyors left. The farmers cheered.

The next morning the crew came out again and had even less success. This time the farmers would not permit them to remove their equipment from the truck. Spokemen Don Jacobson of UPA and Bob Weiss of CPA came to their aid. Jacobson, a tape recorder in his hand, turned to Sheriff Rogers and demanded he move the farmers. It was a tense situation. Several times he asked Rogers, "Are you willing to do your duty?" Rogers responded that he did not want any confrontation. John Tripp remembers what happened next: "Jacobson said, 'That's all I want to know,' and they left."

A strategy session was held by UPA management that afternoon "in an attempt to devise a method whereby the situation could be bypassed." The following morning a large dump truck appeared with survey equipment mounted in the back. What the farmers and Sheriff Rogers noticed right away was that the burly men in the dump truck wearing hard hats were not the usual survey crew. Rogers sensed trouble. He immediately called forty police officers from adjacent counties and cities, as well as the Minnesota state highway patrol.

The men in the truck aimed their surveying equipment over the heads of the crowd, but some farmers raised large placards on poles to block their view. Leonard Lien, a man in his early sixties, carried a placard which read, *Keep Health Hazards Out.* Someone yelled, "Get that man with the sign."

One man leaped from the truck, jumped on Lien, broke his sign, and pushed him into a ditch. Sheriff Rogers grabbed the man and tried to wrestle him to the ground. Then Ben Herickoff joined the fight. Herickoff, a sprightly seventy-one-year-old who is only a little over five feet tall, recalls: "I was just kind of walking along and seeing what happens, and one farmer, he had a big sign and stood in front of the tripod, and then all hell broke loose. Some young guy jumped from the big truck and grabbed that farmer's sign and broke it in pieces and he hollered for help and then the sheriff was wrestling with that big guy. My blood was cooking so I helped the sheriff. I jumped right in and helped the sheriff and I kicked that big guy's fanny—he was lying right on top of the sheriff. And there comes another sheriff, a young sheriff, a deputy, and arrested me. . . . I could never understand that. I always heard, if a sheriff or cop is in trouble, say on the streets, and I am a layman, I am supposed to jump in and help him. And there I got arrested."

Moments later another man from the truck broke another farmer's sign. There was more pushing, shoving, and wrestling. The two men from the truck and Ben Herickoff were arrested and jailed. Leonard Lien received leg injuries and was taken by ambulance to a hospital.

At a meeting later that afternoon in the Meeker County courthouse, Tony Rude, the manager of electrical operations for UPA, told the protesting farmers and news media, "I can assure you that the board of directors of UPA has said the surveying will be done. If it requires violence to get that done, that's what it's going to take."

On October 4, district judge John C. Lindstrom, responding to the violence, issued a temporary restraining order prohibiting surveying in Meeker County. It turned out to be only temporary. Two weeks later, he reversed himself with a permanent order prohibiting anyone from interfering with the surveying work.

In his opinion, Judge Lindstrom noted: "The legislature in enacting the Power Plant Siting Act provided numerous opportunities for citizen participation in the entire process. The plaintiffs did elect to participate and expressed their views not only before the Environmental Quality Council, but also in the proceedings before the Minnesota Energy Agency in granting the certificate of need . . . the plaintiffs have had ample opportunity in these public hearings spanning a period of 18 months before the various administrative agencies to express their opinions." He concluded: "The defendant utilities have elected to submit themselves to the orderly procedures and processes set forth in the Power Plant Siting Act. They have complied with the requirements of the law. The rights which they have established under the law now need the protection of the law."

That decision ended the protest in Meeker County. The farmers did not want to be arrested. The surveyors were able to complete their work.

Pope County presented a more serious challenge for the power companies. A new and more militant leadership had developed near Lowry where the line had been moved by the EQC. These farmers were irate. They had not taken the EQC hearings seriously since the line originally was not routed over their land. They felt betrayed by both the EQC and southern Pope County farmers like Harold Hagen, president of CURE, who were no longer under the line.

The anger that had been building exploded when the surveying crews came onto their land. For five frigid November days they made their stand on a small hill near Scott Jenks' farm outside the town of Lowry. They named it "Constitution Hill." Jenks recalled: "It was a battle to protect our land and our rights. If they couldn't survey past here, they couldn't condemn the land. That was our thought. And they had to stand on top of this hill to survey to the south and to the north."

On Friday, November 5, when the surveyors reached Constitution Hill, they were met by a group of farmers. Lorraine Jenks said: "There were fifty people the first day. They just saw the surveyors on this hill and I think everybody had the idea this was the place they could stop it."

Scott Jenks confronted Al Kingsley, UPA field representative: "You just force your way through here. You don't care who you hurt. You don't care about nobody but the power people." Kingsley admonished him, "I don't think civil disobedience will solve the problem." Jenks replied angrily, "Don't you think the line is civil disobedience to us."

Jenks then drove his truck in front of the surveyors: "I put my truck, my big truck, right in front of them and the sheriff came out and I just have to sit there and he come over to talk to me—he said, 'You're going to have to move that truck or you'll get a fine.' I said, 'How much is the fine going to be?' He said, 'It will be three hundred dollars and/or ninety days in jail.' So I stood right in the middle of the road and I asked those guys [the other farmers],

'Should I move my truck or get ninety days and a three hundred dollar fine?' One guy said, 'Just move your truck.' So I drove it away and he pulled his in right behind and you see, that went on. Then the sheriff had to go to that guy. Dennis Rutledge was the first one. And then Dennis Rutledge pulled his out and another pulled in, and it took up a lot of time and pretty soon the power people just pulled up and left."

The following Monday was "chainsaw day." Scott Jenks: "I took my chainsaw up there one day and I knew it would make a racket in their walkie-talkie. I stood on my land and just run my chainsaw. They [the surveyors] could see right down there, but they couldn't talk. That chainsaw made such a racket on their walkie-talkies, they didn't know what to do. They just stopped everything then and called that boss— Don Jacobson. It took him quite awhile to get here. Then he came out here so we had to start the chainsaw for him. He looked around, but didn't know what to do. Neither did the sheriff. So they left."

On Tuesday a different tactic was employed. The farmers asked permission from their town board to repair the road leading up to the survey site. When the surveyors arrived early in the morning, Scott Jenks was already out on the road working with his digger. To "assure his safety" the farmers barricaded the road. Several hundred farmers came to Constitution Hill that day to back Jenks.

The protesters were well organized by then. They alerted and mobilized their neighbors by telephone and CB radio. The media were also helpful. Every morning local radio stations carried news of when and where the farmers would meet for the protest. Each day the crowds were larger and the protest grew more intense. In the face of this opposition, the power companies called off their surveying in Pope County.

Two weeks later, armed with a court order, the surveyors tried to resume work. Judge Lindstrom had issued an order requiring law enforcement protection for them, so Pope County sheriff Ira Emmons and his two deputies escorted them back to Constitution Hill.

But they were stopped again. One farmer after another stood in front of the surveyors with protest signs. No sooner had the sheriff and his deputies read the judge's order to one farmer, than another would take his place.

Sheriff Emmons decided to read the order to the farmers on the loudspeaker on his police car. He told them that was their last warning. But they refused to listen. Instead they moved out into the field and sang songs—"America the Beautiful," "The Star-Spangled Banner," and "We Shall Overcome." Then they blocked the surveyors again, claiming they never heard the judge's order.

Sheriff Emmons did arrest five farmers that day, the first to be arrested in Pope County, but he was unable to quell the civil disobedience. Sheriff Emmons said: "It wasn't so much that the people were violent. It was the idea that they was getting in the way and the work was stopped. All we could do was tell the workers to quit and move on until the protesters were out of the way."

Sheriff Emmons later described the position he was in: "I had a direct order from the court which said that I should enforce this court injunction. It don't give you any choice; when you are in law enforcement, you are put in here to enforce the law and whether you like it or you don't like it, you enforce it."

Personally, however, he sympathized with the farmers: "I think that the feeling of most of the people out here that were doing the protesting was that was the only thing they had left that they could do. Not that they wanted to. I'm sure a lot of them didn't want to, but they felt that was the only way they could protect their property . . . maybe it was the only way they could get the attention of the legislature and the state, too. . . . My own personal opinion, for what it's worth: I kind of felt sorry for these people, where the law did give the power companies the right to go ahead and condemn their property and take what land they wanted to without the farmer having a whole lot to say about it. . . ."

Jim Nelson and Harold Hagen, concerned about the possibility of real violence, called governor Wendell Anderson and asked him to meet with the farmers. Initially he refused. Then they called lieutenant governor Rudy Perpich, who was to become governor in January; Wendell Anderson was on his way to replace vice president-elect Walter Mondale as senator from Minnesota. Perpich promised to meet with them "the day after I'm governor."

Later Monday afternoon Wendell Anderson called Harold

Hagen and requested that he bring four or five farmers to the governor's office in St. Paul the next morning. Virgil Fuchs, John Tripp, Jim Nelson, Russell Schmidt, and Harold Hagen attended the meeting. They asked the governor to call a moratorium on the surveying. He said he had no authority to declare a moratorium. Instead, he suggested they discuss "reasonable proposals" dealing with new forms of compensation the farmers might receive for the line.

John Tripp remembers that the meeting did not go well: "We got to Governor Anderson's office, we had this meeting, and he didn't want to talk about the problem. He talked about how hard it was to find a house in Washington, how he didn't want to go to Washington in the first place; he hated to leave Minnesota and houses cost so much out there. He was gonna come back to Minnesota every weekend, because he said, 'I am prepared to support an improved powerline law—that's an important thing—if I back it the legislature will probably approve it, and that's a big thing and an important thing and you people ought to accept that. I will be coming back to Minnesota every weekend and I will make sure that it gets passed by the legislature.' He said, 'Tell me what you want.' And we said, 'Stop the line.' "

The impasse between Governor Anderson and the farmers put Pope County sheriff Emmons in a difficult position. The power companies made it clear they intended to proceed with surveying and the farmers, several hundred of them, made it equally clear they intended to stop them. Emmons asked the Pope County commissioners to make a formal request to Governor Anderson for National Guard assistance, but they refused. The commissioners argued it was up to the governor to resolve the law enforcement problem. They voted instead to ask CPA and UPA for a thirty-day moratorium. The companies said no.

On Monday, December 7, Pope County attorney C. David Nelson asked Judge Lindstrom to request National Guard assistance, but Lindstrom declined. Instead he ordered Sheriff Emmons to protect the surveyors. The judge instructed the sheriff to hire up to one hundred deputies at seventy-five dollars a day if that was what it took to protect the survey crews.

The sheriff was not happy about that. He told Judge Lindstrom that he would only hire people who could handle a riot situation, that he only had three nightsticks in the office and very few squad cars, so he wasn't sure how much good the additional manpower would do. The county commissioners wanted to know where the money would come from to hire the additional deputies. They agreed that Pope County was not going to pay seventy-five hundred dollars a day for that purpose. As it turned out, they had nothing to fear. After several weeks Sheriff Emmons had found only three men that were "qualified."

The Pope County sheriff would not hire more deputies. The county commissioners would not pay for additional manpower and refused to request National Guard assistance. As his last act as governor, Wendell Anderson was not about to send in the National Guard against Minnesota farmers. As the Christmas holidays and the New Year approached, the surveyors were unable to continue.

In 1976, the farmers had moved from the hearing rooms of state agencies, to the courtrooms, to the fields. They violated agency rulings and judicial decrees and stopped the line with massive civil disobedience. They had not yet lost their legal fight; there were several court cases pending. But the courts did not appear to be their best hope. What had worked was militant protest.

For officials of Cooperative Power Association and United Power Association, 1976 was a frustrating year. First they were forced by local opposition to go through a time-consuming state hearing process. Then, even with the backing of the Energy Agency, the EQC, and several court rulings, they still were unable to proceed with the line. Once again, they were stopped in Pope County.

Virgil Fuchs
Profile

Virgil Fuchs, his wife Jane, and their five children own and work two 160-acre farms south of Elrosa in Stearns County. When the powerline route was first moved north of the Bonanza Valley, it was slated to pass over the farm where they lived. When it was moved again by the state routing process, it landed on their other farm. Virgil is quiet and self-assured, a man of deep convictions about a farmer's obligation to his land. His quiet determination, his political acumen, and his commitment to action, if necessary, made him a natural leader of the powerline protest.

How long have you been living here?

Since 1961. We got married in '61, but I have been farming it since '58. This here is a 160-acre farm. I've been working the land from the time I was sixteen. It is a dairy and crop farm. My dad first bought it in '59 and sold it to us in '65.

Have you been farming full-time since you were sixteen?

Well, I used to also sell snowmobiles, from 1969 to 1974.

Why did you get into that line of work?

I always was mechanical. Well, I looked at it—all these years of working. I said it is just about time when a guy instead of working at the same old thing of farming every day ought to get out and meet people. But after a while it became more of a business than a hobby and I didn't like that any more.

Has farming become more difficult?

People can't take the pride in farming like they used to any more. It is getting very hard. Then the farms are going to be corporations and then they don't care about the land any more. This is what is happening. The corporations are coming.

Anne Fuchs [Virgil's mother]: It is too hard for young couples to start.

Jane Fuchs: Now you can hardly buy small machinery. It is engineered for big farms.

Anne Fuchs: Then there is such a big price on it.

Virgil: It takes at least one-quarter of a million dollars to start up and that's the minimum. It costs a thousand dollars an acre, much less the money you need to start a herd of cows.

Anne Fuchs: Now the inheritance is so much you can't even try and give the land to your children a little cheaper. You hate to give a child a start and then have him lose it. It's a problem. The future looks really rough for farming.

Virgil: A lot of the people that didn't like this line are seeing other pinches that are coming along, too. They are looking at the oil companies. Everybody is taking a price out of the farmer. The price of grain is bad and fuel is going up. Nobody likes the energy people any more.

If other people wanted to know what worked, what didn't work, and why, for the protest, what would you say about working through the courts?

You'd have to try the courts, but I know in our case it didn't make any difference what happened in the courts. I still think, and I probably shouldn't after everything that has happened, I still think that we should be able to win this thing through the court system that is there, but it just doesn't seem to be working for us. I don't know, I guess when we were in the courts or were going to the public hearings, we were trying to work within the system. But looking back, there was no way the system could work for us. There was no question of a powerline or no powerline, it was just a question of where to put it.

How did you look at the public hearings at the beginning? Were you optimistic?

At the beginning they were really a great thing. George Durfee at the EQC said, "You come to those public hearings and you'll have your input and they'll listen to you." Right now, the way I look upon public hearings, we should have went in there and just raised holy hell. That's what the people should have done, I guess.

How would you evaluate the role of the state government?

I am sure a lot of farmers now would say the state was in cahoots with the power companies. I would say that it is a little bit of each. Initially, the agencies had no way of judging and the co-ops deceived them. Then they made a decision and they have to stand behind the co-ops now.

Some of the farmers have told us the co-ops offered to move the line off your farm. Is that true?

This was at the Glenwood public hearing and at that time I should have taken them up. But there again, having faith in the public hearing process, I wanted to present the evidence to the lawyers under oath at the hearing. But I didn't know the things were being decided off the record. See, this was a hearing in recess and one of the attorneys from the power companies came around and said, "Just show me where your place is." I showed him, but I didn't show him where my dad's place was, or my brother's, or our other place. I figured it to be almost like unfair, unethical, for me to do this outside the public hearing—that I was conspiring against somebody else if I had told him at that time I got this place and this place. Now I realize it was a stupid thing to do. I should probably have looked for ourselves. Now you realize that. But I still felt I would have just been moving it onto someone else. They did, you know, move the line off the farm we live on, but they moved it right across our other farm that we work.

Do you regret having been involved as a major leader in the protest, given all that has happened? If you had to do it over again, would you have stayed out of it?

No. If, as you say, I could do it over again, I would have went to one damn public hearing and sent into the public record telling them that you put that line across our farm and I am going to take care of every last one of you guys. That is what one old guy did— seventy-two years old—he says, "You come across my place and I will use this shotgun," and they didn't go across his place either.

You think there are serious health and safety problems?

Oh, yes. At first when I started this thing here I wouldn't have believed that the government would do anything to the people that would have put their health and safety on the line. But after going through these public hearings, I learned that the state indeed did know that there were health and safety questions associated with the line. They used them as criteria in routing this line. Only we couldn't use the health and safety because we couldn't document the problems.

Do you think there is any chance this line can be stopped?

Oh, yeah. I guess I'll believe that for the next twenty years. This line should have been stopped. I was real confident the supreme court would make the power companies rehold those hearings where there was missing transcripts, and then I know what would have happened at those two hearings. At that point we would have learned and we would have raised holy hell. I think the state knew that. Right now the state of Minnesota wouldn't let the EQC people take their cars out in the country anymore. A person doesn't feel too safe driving a car with the state emblem on it.

Do you think what has happened out here has mainly just been self-interest?

To a certain degree it might be. I don't think there would be many people here who would pick themselves up and go to Wyoming to protest stripmining. I would say that there would be more to do it now than three years ago. Now there are certain ones who are more interested in what is happening as a whole, than they would have been before they went through this. See, the system will destroy the system if this continues. Another ten years you get these people affected by this, you get people affected by other things—coal mining or nuclear plants, and it is just like a mushrooming with their families and their children. Somebody says, "Well, why are you protesting the nuclear plant?," and you don't really know why, but there is some reason somewheres along the line when you were a little kid or something you heard something happened to somebody else that you didn't like what the system did to these people. And that is why you are against something the system is trying to jam down somebody else's throat. Like these little kids here—who knows what they're going to turn out to be? I would suspect that they will probably be a lot stronger on things like what is happening to the people when they get to that age because they have gone through it in childhood.

Ben Herickoff

Profile

After immigrating to the United States from Germany in 1930, Ben Herickoff built an initial few acres into the most successful farming operation in Stearns County. Retired now, he and his wife live in Elrosa while their sons run the farm. At one point the powerline was routed across his farmland; later it was moved, but Herickoff remained on the front lines of the protest.

You testified at a public hearing about back in 1936 and '37 when you'd been going around on your bicycle to get the REA started.

Yessir, I worked just as hard that time for as I do now against this high powerline. This REA is a fine thing; it's the most finest thing could happen to us, but now it's get to be overdone. We don't like to be killed by the line. We'd still like to live a few years yet.

We just wanted to get a feeling for what it was like back then. How did you get involved with REA?

Well, I knowed something about electricity. As a rule, in Germany, with a lot of things they're behind, but on electricity they were ahead of us. We had it over there already, on my dad's farm, and here we didn't have that yet till when this REA came up to our local people here. Most of them are dead already. Anyway, we went out, going from farm to farm and trying to sell shares and we had ever slow going. For example, we had some people, some of my neighbors that said, "Ben, why would you push such a thing like that; electricity is dangerous." I said, "I don't think so that it's dangerous; naturally, it's dangerous if you don't watch it and if you don't do it right."

You were brought up on a farm in Germany?

You betcha.

How old were you when you came to the United States?

Twenty-three.

What year was that?

1930.

Did you settle right in this area?

No, I came first to Freeport [Minnesota]. I had some friends and relatives down there, so that's why I got over there. Then I learned to know my present wife.

When were you married?

1931.

And then you moved up here to Elrosa?

Yessir.

In what year?

In 1931 we moved up here. The first year I worked for a farmer; I got three hundred dollars a year working there, and I saved that money and then the next year I used that three hundred dollars as a down payment to get my farm. It was cheaper making a down payment than renting. Renting a farm was high-priced that time, and so we bought a farm. Naturally, the old buildings on it were in bad shape, but the land in itself is good. When we were going there I was one of the lowest taxpayers because the land was unfertile, more than half of it in water—no ditching or nothing. Another twenty years later I was the highest taxpayer in the county.

Could you tell us a little bit about how that happened?

Oh, real simple. When you improve, you work hard, you do

something good, usually you get fined. That's what it was, see. We drained all the land and cleaned it up, got the weeds out of the fields and do the right kind of job of fertilizing and things like that and so it didn't take too long, we came on top. Nowadays they say, "Yes, that Ben Herickoff, he can make it"; that time, they wouldn't give Ben Herickoff no credit for a couple of years—they thought he's never gonna make it.

You said you thought the REA was really a great thing. Why did you consider it a great thing?

Well, everything is so handy and you had the lights. Before we went with oil-burning lanterns, and pumping water by hand, and then we just put a switch on and with a motor and a pump water comes.

You were one of the first ones to want the REA?

Yeah, and pretty soon everybody wants to have it. At first they kind of worked against it like I was saying, but that was the right thing for us, I'll say. I thought a good thing. Till now when they wanta build this high-power line. I can't see it.

What's your main concern about the line?

Oh, why I am against it? Health hazard, mostly. The health hazard to me is my big objection. At the meetings we had, the health question came up, and the powerline guys, they always denied that there was a health hazard, and it just happened right here in Elrosa that we had a meeting and some of their lawyers said, "If you don't believe us why don't you go to Ottawa, Canada? There's an experiment station there and you will find out how good it is, how nice it is." I wrote it down, and when I came home I said to my wife, "Wake me up at four o'clock—I want to be leaving early to be at the airport." I had very good luck. It was foggy like hell, and then she didn't want to let me go either. I had to drive thirty-five miles an hour and it took four hours. But I got a plane standby and by noon I was in Ottawa.

That very night you decided to go?

Yah, just by myself. I like to see the truth come out on top; doesn't make much difference, whatever it is. I couldn't say much because I was just a good listener. But my boy was talking against this, too. He's a doctor in science; he's teaching at Mankato. So this kind of convinced me against it, but I wasn't sure yet. But them power company guys, they thought there wasn't nobody gonna go see—a thousand miles or more—nobody will go anyway, dumb farmers, they don't do that anyway.

Did you know the name of the place that you were going to?

No, just they said the experiment station in Ottawa. Well, I thought, that's enough. As long as you have your mouth with you, you find out things, the same as you hear. You never would have found me except through the mouth and good hearing and things like that, right?

Who runs this experimental station? Do the utilities support it?

Oh, it's supported by the government of Canada.

What happened when you got there?

Naturally, I asked for the professor who's the head man. I couldn't talk to him; he wasn't in that day. So I met the next higher-up, who was an engineer. When we got to start talking, he said, "By golly, Ben, you must be a foreigner." "You bet," I said, "just like you, I believe." I could hear that on him, too. He said, "From where are you?" "Well," I said, "I was from Germany," and I named the state. He said, "Then we are just about neighbor boys." He was from Holland and I was from Germany, but we both lived on the line. So I was all afternoon with him. He took me out for coffee, for supper, and he seen that I had a reservation for a motel, things like that, and then we made a reservation next morning that I could meet with Dr. Morris.

Did they have a DC or AC line at the station?

DC. They had a converter there. It was just a short line, about two hundred feet long. They converted it from AC to DC. When it came up to four hundred [kilovolts], it started a little bit crackling already but not just really too much. Then when they went up to six hundred, seven hundred it got real loud already, and when they got to eight hundred, it felt like a thousand needles sticking through my head. So I said, "Just a minute, I'll go back and put my hat on." Came back; it didn't mean nothing. I had the same sensation. This Dr. Morris, he said, "Do you know that your hair is standing up?" And then some of the other guys, one of them an engineer, he thought he must have a little fun. He came by me and shaked hands with me while I was under that line there, and whoa, didn't I get a shock. He said, "By golly, you're really loaded with electricity." That's where my rubber boots came in. So then they started explaining that the same thing will happen if you stand under the line with a new combine, with good rubbers on it. That thing will load it with electricity and you come there and take hold of it, you might be a dead duck, he said.

When was your trip to Ottawa?

In March of 1975.

Did you come back and report on it?

Yes, at the meetings, yessir.

Did you ever tell the power companies about your trip?

Well naturally, at them big public meetings.

What did they say?

Oh, that couldn't be so. They tried to make a liar out of me. They said, "It isn't so, don't believe what that Herickoff said, don't believe that." Things like that was said at the meetings, and that made me more mad. It is surprising how mad you get, especially when you're saying the truth.

When did you retire from the farm?

'74. I was off the farm already. The boys were farming. But still the farm was in my name.

Is this the same farm you had in 1931?

Yes, it's the same farm.

How many acres did you start with?

A hundred and sixty.

And you got up to twenty-three hundred we hear?

Yes, just about. Not just me. The boys and I. We bought more land and things like that. As things went on, we went dairying, we had purebred Holstein cows, we had a lot of good luck, and naturally it took all of our effort, too. It didn't come by itself. We stayed home quite a bit; we didn't do much loafing and running and chasing around business.

You must have worked very hard. Are you glad you're a farmer?

Farming is a quiet life. You don't get interfered with by no neighbors; with hardly nobody, except the government.

Is it true today that a lot of corporate farms are taking over from family farmers?

I'm not acquainted on that. I don't like to see corporations so much. Sometimes it's good, but the worst thing is when it goes into ownership and the corporations, they make a dictatorship out of it. And that's what it is with REA; the big shots must have thought there was good money in it. And they were all better educated than farmers. So slowly the farmers, they give the rights up and these big corporations took over.

How did the farmers give their rights up?

Well, too busy, not enough educated. I daresay nowadays every farm boy went to high school, a lot of them went to college. All of our kids went to high school and the biggest part of them went to college, just about all of them, but not in them early years.

How many children do you have?
Mrs. Herickoff: Ten.

How many of them have stayed in farming?

Two boys and three girls. Some of them doing better than dad!

When you came over from Germany, did you come over by yourself?

Yes, I came by myself. I was twenty-three.

Why did you leave Germany?

There was no elbow room. It was populated so big and hardly any jobs to get, good jobs. I used to work in a coal mine, dug coal about three thousand feet deep, more or less.

You were a coal miner?

Yessir. I sure worked the ground, but down *in* the ground; I would sooner work above the ground.

When did you start working in the coal mines?

As soon as I could get in, when I was eighteen years.

Was there no place to farm? Didn't you grow up on a farm?

Well, there were three brothers; the oldest naturally gets the farm. It's just like that in Europe. The oldest is the next king—if he knows something or he don't.

So you had to work in the mines. What was the work like?

Sometimes you work on your knees, with air hammers you work, or shovels, laying tracks, do some building, make what you call driveways underneath the ground.

Were the wages good?

Well, I thought so—about half a dollar a day.

But you came to this country to farm?

Yes, I desperately wanted to farm.
Mrs. Herickoff: He always wanted to come to America, and that's why he worked; he saved his money so he could come.

Did you speak English well then?

No more as our cat can.

That must have been difficult.

Mrs. Herickoff: He said when he came to New York, it was just like a bunch of geese talking, that's all he could understand, nothing.
I had train tickets. You come from the boat and they kinda lead you to a train. I was on the water for fourteen days and then I

traveled by train to Freeport. There were people living in Freeport that were from the same town, see.

Did you get back to Germany soon after?

It took about twenty-five years, after we were pretty much established. Now, we've been back seven times.

There are a lot of German-speaking people anyway in Freeport. This whole Stearns County is more or less German-speaking. It's turning mostly over now into English.

You were arrested. How did you feel when that happened?

Even though I was never arrested in my life, I felt I was 100 percent in the right and I was not afraid for it; I was not ashamed of it either. I was kind of proud of it. And I think the arrests made more people strong and angry.

Whatever happened to you? Were you convicted?

The charges were dropped, but they took me in the court first and then they asked me, "You plead guilty?" "Hell no," I said. And the next day I had to come back and meet the judge. And he told me what I might have to pay. I said, "For that, we gonna have a jury trial." "We'll let you know when," he said. Nothing ever happened.

You mentioned the public hearings earlier. What was the purpose of these hearings?

Well, trying to still convince the people and telling them a bunch of lies. But people got to be more educated from one meeting to the other and more aggravated.

Sometimes you read in the papers that the farmers out here, because of the powerline, are getting radical. Would you call yourself a radical now, after all this has happened?

They're calling the farmer radical, you're saying? No, they're badly mistaken. As a rule, a farmer is the most peaceful man on earth, much more than anybody else, but he likes to be left alone. I was a salesman at one time, I sold dairy equipment, and I still say the farmers are nice if you treat them nice. You're not going to charge in the house; you ring the bell first, like you did, and try to be nice, you haven't got no troubles with the farmers.

Some people say that if the power companies had come around and explained what the line was about and done a better public relations job, they would have gotten the line through. Do you think that's true?

It would have helped it. That's pretty near like what I'm saying, I'm just saying it in a little different way. Any time if you want to do business with the next man, then don't come with the axe, leave the knife outside. Just with a handshake and being decent. The next man, he has a soul and he has a body, he has a mind, just like you have, like I have, like everybody else has.

If the utilities did that, do you think the farmers would have protested anyway?

If they would have used much more common sense, it would have helped, but I still would be against it—this high voltage line —it is the worst thing we could have.

Rudy, we know what you do,
And we're losin' all respect for you
You know we're gonna see this struggle through
'Cause it's a question of who controls who!

—from "We Don't Want No Powerlines
Around Here" by Russell Packard

5 A Populist Governor

Soon after Rudy Perpich became governor on December 29, 1976, he made headlines by disappearing mysteriously from his office for two days. When he returned on January 11, he disclosed that in an attempt to avert serious trouble brewing over the powerline, he had driven out to west-central Minnesota to meet with the protesting farmers. From the beginning of his tenure in office, Perpich personally committed himself to resolving the powerline controversy and publicly proclaimed it a high priority of his administration.

The governor had visited Pope County. While driving home from a convention in western Minnesota to nominate a successor to representative Bob Bergland, who was to become secretary of agriculture, Perpich decided to find out personally about the powerline. Evading his two bodyguards on the highway, he pulled into the local bar at Lowry and asked how to get to Dennis Rutledge's farm. Rutledge, active in the protest and local DFL politics, remembers this evening well: "It was Sunday night. I was home early from the Reserves because of a snowstorm. Jerry from the bar calls me and says, 'There is a guy all dressed up in a suit who is asking for directions to the Dennis Rutledge farm. What do you want me to do?' I said, 'Ask who it is.' He came back to the phone and said, 'He says, I'm Governor Perpich.' I said, 'I don't know who is playing a joke but send him out so I can see who this joker is.' Pretty soon there was a knock at our door. My wife Nina answered and she shouts out, 'It really is Governor Perpich!' "

Perpich talked with the Rutledges for about an hour and then they went out and met with the Rutledges' neighbors. The governor drove the 150 miles back to St. Paul late that evening only to return the next morning to meet with more Pope County farmers throughout the day.

Rudy Perpich never promised he would stop the line. Mostly he listened to the farmers. He assured them that from his own experience growing up on the Iron Range of northern Minnesota and dealing with big mining companies, he knew how they felt dealing with big utilities and that he was determined to find a solution to the dispute. According to Dennis Rutledge, "He didn't make any rash promises. He just said he felt there had to be a solution. He talked about moving the line along the interstate highway 94 or further north along scrub land not good for agriculture. He also talked about health and safety problems."

The farmers were pleased by Governor Perpich's visit. They felt that Perpich, unlike his predecessor Anderson, cared about their problems. His political instincts were populist; he was not just another politician. His personal involvement raised the farmers' hopes. If any politician would help them, Rudy Perpich was the one.

There were other indications that Perpich genuinely wanted to help. Prior to his mystery trip, in his first "state-of-the-state" address to the legislature, Perpich announced that at his urging House and Senate leaders had agreed to appoint a joint committee to hold hearings in St. Cloud on the powerline controversy.

On Wednesday, January 12, 1977, the powerline controversy reached the legislature. Five hundred farmers filled the St. Cloud armory for a one-day hearing before a special joint committee of legislative leaders. The farmers asked the legislators to overhaul the EQC, to change eminent domain laws, to give higher priority to farmland in siting energy facilities, to promote energy conservation, and, most important, to reconsider the CU powerline.

Jim Nelson captured the prevailing sentiment and received sustained applause when he told the committee that farmers were frustrated and angry with the treatment they had received from state agencies and that the legislature should order the agencies to conduct their hearings over again. Nelson also called for a one-year delay on the line until health and safety questions were carefully studied. However, Gerald Willet, chairman of the Senate agriculture and natural resources committee, replied that any attempt by the legislature to pass a moratorium bill would invite lawsuits from the power companies. He and the other legislators seemed more interested in legislation concerning future power lines.

Phil Martin, general manager of UPA, left little room for doubt about where his company stood on the Pope County protest and these first legislative hearings. He told the *St. Cloud Times,* "People keep asking me what I'm going to do now. Do you want me to say I'm going to hire my own army? Do you expect me to roll over and play dead? I expect the law to be carried out. I'm going to build a transmission line."

Governor Perpich invited some of the protest leaders to his office a few days after the legislative hearing. Alec Olson, the lieutenant governor, did most of the talking at this meeting on January 18. He gave the farmers reason to believe that under the Perpich administration things would be different. According to Lorraine Jenks' notes from the meeting, Olson said the line was not a foregone conclusion: "There is no such thing as 'this line has to be built.' We can start over." He spoke of alternative routes such as railroad rights-of-way and how important it was to address the question of whether or not the line was needed. He told the farmers he wanted, in his capacity as the governor's official representative on this issue, to get negotiations going between protest leaders and power company officials.

The farmers left this meeting with the euphoric feeling that they had a real chance to stop the line. John Tripp recalls, "Gosh, I remember that meeting well. We were on top of the world."

At the urging of Governor Perpich, about twenty-five leaders met in Glenwood the next week with power company officials. The farmers demanded that the line be built somewhere else, if at all, and that a moratorium be declared until health and safety effects could be studied. The companies maintained that the proposed alternatives, such as building the line along interstate 94, were not feasible. The only thing they were willing to negotiate was easement payments. As far as they were concerned, the route was fixed; there would be no changes or delay.

A real blow to the farmers' hopes was a letter they received the following week from Ronnie Brooks, special assistant to the governor. Brooks outlined a proposal "aimed at generating productive negotiations between the representatives of the power coopera-

tives and persons directly affected by the line." The key paragraph
read: "The proposal is based upon recognition by the Governor of
the fact that additional effort is required to reach a resolution of
this issue and that a complete negation of the present route and
the initiation of an entirely new siting process for the CPA-UPA
transmission line is not feasible." She proposed instead that the
state obtain the service of a professional facilitator to help medi-
ate the dispute.

Only two weeks before the farmers thought they had been told
the governor was not committed to the CU project, that everything was negotiable, that he intended to reopen questions concerning need, health and safety, and alternative routes. They were
incensed by Brooks' letter.

This was the beginning of a bitter relationship between the
farmers and Ronnie Brooks, an activist liberal in Minnesota politics, who had risen to political prominence directing the Minnesota McGovern presidential campaign in 1972 and who, as time
went on, assumed more and more responsibility for handling the
controversy. She has reflected on what happened from her perspective: "Rudy Perpich was higher than a kite after his first visit
out to western Minnesota. He felt he could solve anything. Then
he realized what a quagmire it was, that undoing and redoing was
much harder than doing, and he started backing off. . . . He wanted
to get rid of it. My job was not for the farmers or the cooperatives.
My job was to save the governor's ass. No politician likes conflict.
I became exceedingly political. I was sucked into the role."

Governor Perpich announced that he would personally bring a
professional mediator out to west-central Minnesota to introduce
him to the farmers. On Wednesday, February 9, hundreds of farmers gathered at the town hall in Lowry (population 257) to meet
Josh Stulberg. Stulberg, thirty-one-year-old New York lawyer,
national director of the community disputes services division of
the American Arbitration Association, explained that he had
come to Minnesota to talk with the farmers and power company
officials to determine whether he could play a positive role in
mediating the powerline conflict. He emphasized that both sides
had to decide if they wanted him as a mediator; if so, the first
thing he would do was "keep the lines of communication open,"
in order to address the genuinely and deeply held concerns of

groups and individuals. In an apparent reversal of the position in Brooks' letter, Stulberg and Perpich said the negotiations would deal with a wide range of issues and listened as the farmers called for reconsideration of alternative routes, health and safety problems, energy conservation, and the question of need.

For the next two weeks, Stulberg met "informally" with farmers and company officials. At the last of his preliminary meetings with the farmers, held in Alexandria on February 24, Stulberg stressed the need to establish an area "for productive discussion that can open up the lines of communication. It doesn't take a genius to realize that nobody trusts anybody." Virgil Fuchs responded: "We're not hiding anything. We told them—we're going to stop you guys if you want to come across our land. They can trust us. We haven't told a lie." When Stulberg asked, "Are there ways for addressing the issues?," Alice Tripp voiced the fear of many of the farmers: "We've got them stopped—since November 1. Are you sure we're not negotiating a surrender—because right now we're on top."

The meeting concluded with Stulberg's asking, "I guess I would like to have some feel from you whether it is worthwhile my returning?" The farmers were agreeable. But John Tripp cautioned, "We're not going to talk about where the line is going to be located."

The much publicized mediation talks among Stulberg, power company officials, and the farmers were held March 16 at the Americana Inn in St. Cloud. It turned out to be the first and last session. Stulberg began by asking participants to "identify items for future discussion." He warned that "the atmosphere in western Minnesota is potentially dangerous" and urged both parties to negotiate in good faith.

Immediately the meeting went sour. One issue was tape recorders. Stulberg asked that all tape recorders be shut off; he felt they inhibited discussion. The farmers refused; they said the recorders were necessary to give the other protesters a complete and accurate account of what happened at the meeting. Stulberg asked for assurance that the tapes would only be used for that purpose. Virgil Fuchs replied, "I feel it is my private property and I will use it any way I want to. The only way I can be here is if everything is recorded. It is up to you."

The farmers maintained that the presence of power company lawyers John Drawz and Roger Miller did more to inhibit discussion than electronic devices. In their earlier meetings with Stulberg, they had requested that Drawz and Miller not be present at the mediation sessions. They said both lawyers had "grilled them unmercifully" during the EQC hearings and they did not trust them. They argued that lawyers' fees were no problem for the power companies, but their groups could not afford to hire attorneys for the mediation sessions. If the CPA and UPA lawyers were to be present, they wanted their own lawyers—at state expense.

Stulberg recommended a compromise—the farmers would remove their tape recorders and the lawyers would not talk. The farmers emphatically rejected that idea and the session went on—with tape recorders and lawyers.

When they finally got down to business, Stulberg tried to steer the discussion to health and safety concerns. He indicated that the other issues could better be resolved in the legislature and in the courts. He wanted to discuss the possibility of a new mechanism—something called a "science court"—to deal with health and safety.

At this point, before they really started, the negotiations broke down. At issue was a fundamental question: what would happen while this investigation of health and safety was going on? The farmers insisted that any such action be accompanied by a moratorium on construction of the line. They wanted the companies to promise they would stop all work on the project, including surveying land, obtaining easements, and ordering equipment. They feared a health study would drag on for several years and in the meantime, if there were no moratorium, the line would be completed. They wanted to know how farmers could be expected to go to mediation sessions and discussions of the issues and at the same time be out in the fields stopping surveyors and easement men.

UPA manager Phil Martin and CPA manager Ted Lennick denounced the demand for a moratorium as "blackmail." After the meeting they announced that after four-and-one-half months of "voluntary moratorium" they intended to proceed with surveying and land condemnation along the route.

Stulberg was disappointed. He described the moratorium demand as "a proposal that got in the way of discussing some of the more important issues."

Governor Perpich was displeased. He blamed the farmers for the impasse. He told the news media that the farmers had made a "big mistake" in insisting on a moratorium and "could lose a lot of friends" by their action.

Perpich's stance angered the farmers. Jim Nelson called to tell him that "the power company was causing the impasse, not us, by insisting we go along with the line as if nothing were happening."

With the breakdown in negotiations, the struggle "within the system" shifted to the legislature and the courts.

Powerline siting was a major issue in the 1977 session of the legislature. Many of the farmers traveled frequently to St. Paul during the winter and spring months to talk to legislators and testify before legislative committees.

In late May, the legislature passed a bill that changed the criteria governing the routing of future powerlines. The legislation changed the process by which future powerlines would be sited, eliminating corridor selection and making route selection a one-year process. It altered the criteria to be used in selecting a route, for instance giving agricultural land equal footing with wildlife areas. However, there was nothing at all in the law that pertained directly to the cu powerline.

The farmers were bitterly disappointed by the legislature's action. John Tripp called the bill a "payoff," intended to placate the farmers without addressing their central concern. The farmers, he pointed out, were fighting one particular powerline, the cu line, and the legislature had avoided that controversial issue entirely.

This was the beginning of the farmers' disillusionment with the legislature. They would come down to the capitol, wander around and talk to a few sympathetic legislators, but they remained outsiders in a system dominated by insiders. They brought controversy to an institution that avoided controversy like the plague. Jim Nelson has reflected on their experience: "Although we might be down in St. Paul as often as once a week, the power company lobbyists had the other four days to work on the legislators. They made good use of the other four days." Steffen

Pederson, a sixty-five-year-old Pope County farmer, had a similar reaction: "It seemed like the power companies could maneuver the legislature—it looks like it. I think they have real intelligent lobbyists working with the legislators—they go right in their offices and propose what they need and just get them to think their way. When we came to the legislature, they treated us like strangers." Alice Tripp says she will never forget the time one of the power company lawyers got up at a legislative hearing and began his statement with, "We who worked on this bill. . . ."

The last hope was the courts. For an institution notorious for its slow, drawn-out processes, the response of the judicial system in this case was surprisingly swift. Several protest groups—CURE, No Power Line, Save Our Countryside, Preserve Grant County, and Families Are Concerned Too—had brought suit against UPA, CPA, and EQC, and the Energy Agency. On March 1, the state supreme court consolidated seven suits into one and named a panel of three district court judges to hear the case. Three weeks later the panel convened in Glenwood and ordered a halt to work on the CU line until it had ruled.

The protesters had hired a bevy of lawyers: Norton Hatlie for CURE, George Duranske for NPL, SOC, and PGC, and David Grant and Elani Skevas for FACT. They raised two principal issues:

1. Once UPA and CPA had been grandfathered out, the EQC no longer had jurisdiction over siting the line. The savings clause in the 1973 Power Plant Siting Act (PPSA) explicitly stated that the provisions of the act "shall not apply to . . . high voltage transmission lines, the construction of which will commence prior to July 1, 1974; provided, however, that within 90 days following the date of enactment, the affected utility shall file with the council a written statement identifying such transmission lines, their planned location, and the estimated date for commencement of construction."

Within ninety days following the enactment of the PPSA, UPA and CPA *had* filed the requisite written statement and as a consequence they had been exempted from the state routing process. Since the PPSA stated that under those circumstances the provisions of the act "shall not apply," the farmers contended that the EQC's later acceptance of jurisdiction when the companies ran into trouble in Pope County was a clear violation of the law. FACT asked that the process be returned to the counties: "Because of the history of this case, the reliance of the citizens on the process which existed at the local level, and the intent of the legislature not to force the new legislation on pending projects, remand to the counties to finish their proceedings appears to be the alternative most likely to resolve the existing distrust and ill will."

2. The EQC and Energy Agency had violated the law. Several violations were claimed, including:

A. Irregularities in the corridor and route selection and certificate of need proceedings. Among those cited were: (1) The corridor selection and certificate of need hearings were reversed; (2) The EQC had not prepared a state-wide inventory of powerline corridors and the Energy Agency had not prepared five- and ten-year energy forecasts, both of which were explicitly required by law for the siting and certificate of need processes; (3) No environmental impact statement (EIS) had been prepared at the time of corridor selection; (4) The Energy Agency's failure to include MAPP as a party to the need hearings effectively denied the farmers due process; (5) Larry Hartman's previous association with Commonwealth Associates constituted a conflict of interest.

B. Inadequacy of the environmental impact statement. The EIS that was filed during the route proceedings was inadequate, on the grounds that most of it was copied from the power companies' environmental report and it failed to consider alternatives to the powerline, as required by the Minnesota Environmental Policy Act.

For these reasons, the farmers asked that the proceedings be remanded to the agencies for rehearing.

On the other side was a high-powered lineup of lawyers representing the power companies, the EQC, the Energy Agency, and the state attorney general's office. By now the state had a real stake in the outcome.

On July 14, 1977, the panel announced its decision; the three judges unanimously denied all the farmers' suits. The panel ruled that the EQC and the Energy Agency had correctly interpreted the law; it upheld all of their decisions.

UPA spokesman Don Jacobson said he hoped the panel's decision "lays to rest the concerns expressed by the people." EQC chairman Peter Vanderpoel said he was "gratified by the decision

—from the state's point of view, the line had to go somewhere . . .
I really don't see what could be gained by sending the project back
to do over. . . ." The farmers announced they would appeal the
decision to the state supreme court.

The appeal was filed three weeks later and the supreme court
ordered a halt to work on the powerline pending its decision. It did
not take long. On September 30, the supreme court issued a ruling. It was unanimous—against the farmers.

Could the EQC reassume jurisdiction after the power companies
were once exempted from the provisions of the Power Plant Siting
Act? "Yes," the court decided. Admittedly, there was little guidance from other court cases and from the legislative history of the
act. This was a first in the annals of Minnesota law: "In all the
reported cases, the appellant is fighting to get or retain his
exemption; never has a heretofore-exempt party sought to relin-
quish being excluded from the burdensome regulations enacted
by the legislature." The legislature had not anticipated such a
situation either; as the supreme court opinion put it, "ascertain-
ment of legislature intent is difficult."

In the end, the court said, it based its judgment on the fact that
"the two crucial concepts that permeate the entire act are that the
process should be *orderly* and there should be *public participation.*" Applying these criteria to the case at hand, the court
concluded:

> Centralized decision-making is more orderly than numerous
> duplicative local actions, and all interested citizens were
> given an opportunity to be heard in the public hearings con-
> ducted at each stage of the process. After long and serious
> deliberation, we have decided that the legislature would have
> intended MEQC to have jurisdiction, based on our residual
> conviction that the legislature would have preferred the PPSA
> to apply to the fullest extent reasonably possible. Therefore,
> the legislature must have intended to permit the utilities to
> waive the exception it had carved out in their favor, even
> though exemption had previously been claimed and granted.

The court also rejected all of the farmers' contentions of illegalities and irregularities by the EQC and the Energy Agency. Although the corridor and need hearings had been reversed, "in our

judgment the need for additional electric power was so clear that the order in which the hearings were held was of no practical significance."

True, the required state-wide inventory of powerline corridors and the required long-range energy forecasts had not been prepared, but, in the court's judgment, that was not fatal.

Although "it would have been preferable for the MEQC to have required an environmental impact statement" before the corridor was selected, "its failure to do so does not constitute a reversible error."

"Since MAPP members were never shown to be indispensable to these proceedings, the decision of the hearing officer not to require them to answer interrogatories was correct."

Larry Hartman may have worked for Commonwealth Associates just prior to coming to the EQC, but "the claim of bias—is equally without merit."

As for the environmental impact statement that was finally prepared, the court admitted that the farmers had raised "serious questions about the conclusory nature of much of the EIS and its failure to discuss alternatives to the HVTL. . . ." However, "the mere fact that much of the information in the proponent's environmental statement also appears in the agency's EIS is not sufficient, by itself, to demonstrate its inadequacy . . . we are not persuaded that the EIS was fatally defective."

And so on. The court had a great deal of discretion and it ruled in favor of the agencies on all counts. Irregularities and illegalities were not enough in the case of government agencies: "Decisions of administrative agencies enjoy a presumption of correctness, and deference should be shown by courts to the agencies' expertise. . . ." Thus, the court explained, "to prevail in this appeal," the farmers would have to "demonstrate that the decisions complained of were improperly reached and incorrect, a burden of proof which they have been unable to reach."

In an interview two years later, chief justice Robert Sheran, who wrote the supreme court opinion, elaborated on the fundamental "presumption of correctness" of the agencies: "What the courts are expected to do is to see to it that the administrative tribunals give the parties notice of the hearings, opportunity to be heard, and that they base their findings upon evidence which is presented, made public subject to cross-examination. But to do that in a way so as not to intrude upon the purposes of the administrative agency, which is to bring specialized expertise to bear on the problem at hand . . . we start with the assumption that the administrative agency acted in accordance with law and applied itself deliberately and carefully to the problem at hand. . . . Because of the presumption of regularity on the part of the administrative tribunal, the burden of the challengers is to do more than to show a slight balance of the evidence in their favor. You have to go further than that and show that the decision of the administrative agency was an unreasonable decision. Not only that they chose the view of the evidence which was not supported by a preponderance, but they chose the view of the evidence which a reasonable body would not have chosen. . . . The thought is that the administrative agency has special expertise, that the members of the administrative agency have been chosen because they have special qualifications to deal with the problem at hand and ·the courts are expected to recognize that and give some deference to it. That really is the essence of the process."

One justice disagreed. Lawrence Yetka issued a "special concurring opinion," in which he came down hard on the EQC and the Energy Agency: "In my opinion, what is wrong with this case is that the state agencies involved have misconstrued their intended role in proceedings of this type. They have played a passive rather than an active role." He criticized the certificate of need process, saying "proceedings on the question of need for the facilities were conducted, but they were hastily organized and possible alternative sources of power do not appear to have been seriously considered. . . . It is apparent from the transcripts of the public hearings held throughout the state in connection with the location of the route that the findings of need was a foregone conclusion. . . ." He also criticized the siting process, notably the fixed entry point and the EIS.

Justice Yetka posed what he termed "a very serious question: What should be the proper role of the state agencies empowered to act under this statute? To limit itself to consideration of proposals made by the utilities and any possible counter proposals that might come from the public? Or is it to serve as an independent, impartial arm of all of the citizens of this state to not only take and hear evidence but if necessary to generate evidence of its own?" To him the answer was clear: "I cannot envision a governmental agency being effective in protecting the public without having the authority to itself seek out the facts independently."

Nevertheless, Yetka concurred with the majority ruling against the farmers. He explained why: "I concur with the majority, but only because I can see no purpose in remand except a delay in construction and higher costs as a result thereof. It is obvious to me that under the existing statutes the state agencies will do no more than conduct hearings and approve the certificate of need and the route already selected."

The farmers had pinned their hopes on the supreme court. They were both disappointed and disillusioned.

Jim Nelson recalls his reaction: "I think what got us in the end was basically what Yetka was saying in his decision—that even though the farmers have obviously not been given their full share of rights on this issue, we still have to rule against them." Bruce Paulson puts it more strongly, "I don't think it done a bit of good for farmers to go to the courts; I think we could have just as well saved our money and got violent as hell. If you wanted to stop the line, that's the way to stop it. If it ever happens again . . . there's one thing to do and that's get violent as hell. That's the only way you are going to stop it."

The big question for the farmers was what to do next. The media were declaring the protest dead. Alice Tripp recalls: "After the supreme court decision, the press kind of jumped on the theme, 'We're tired of the story; it's time to quit; what can you do now? How can you fight a supreme court decision? The law has decided.' "

One thing was clear; they were not about to quit. Two days after the decision, more than two hundred farmers met in the Lowry town hall and vowed to keep the powerline off their land despite the high court ruling.

Governor Perpich got a real sense of the farmers' mood when he made another of his unannounced visits to west-central Minnesota a few days after the supreme court decision. His first stop was Virgil and Jane Fuchs' farm. He asked Virgil where the line would cross his land. There wasn't going to be any line, Virgil replied.

While the courts were deliberating, Perpich had made another proposal. On July 7, he announced that the Ford Foundation had provided a start-up grant of $5440 to fund a "science court" experiment to deal with health and safety aspects of the powerline dispute.

The science court concept was developed in 1976 by a White House task force headed by physicist Arthur Kantrowitz. As envisioned by the task force, a science court would have three stages. The first step would be to identify the significant questions of science and technology associated with a controversial public policy issue; the court would be concerned with these questions alone, leaving other questions, political, ethical, and so forth, to subsequent consideration by other elements in the decision-making process. The second step would be an adversary proceeding, presided over by a panel of impartial, objective scientist-judges, in which scientist-advocates would debate the technical questions in dispute; this courtlike setting would feature testimony by scientific experts, with an opportunity for cross-examination. As the final step, the panel of judges would issue its judgment as to the scientific facts pertaining to the disputed technical questions.

The science court idea had been widely discussed in the science and public policy literature and the Kantrowitz task force had proposed an experiment to try it out. However, when Jimmy Carter became president, the task force was abolished before the science court had been tried.

The Ford Foundation grant was awarded to Josh Stulberg's employer, the American Arbitration Association. Governor Perpich asked the two power companies and the anti-powerline groups to participate in a science court, to be coordinated by Stulberg, with a format to be negotiated. By this time, officials of the National Science Foundation had also indicated informally that they would look favorably on a request for a much larger grant to fund the science court once preliminary agreement on a format had been reached.

Stulberg returned to Minnesota in early August to try to convince the farmers and the companies to participate in a science court. The initial reactions he received were not encouraging, however. The farmers did not want the science court to be restricted to health and safety. They wanted it to consider other issues as well, such as need, alternative routes, and eminent domain policy. They reiterated their insistence on a moratorium on construction, arguing it made no sense to build the line while holding a forum on the question of whether it should be built. UPA and CPA officials questioned the need for a science court, pointing out that the farmers had ample opportunity to participate in the public hearing process. They firmly rejected a moratorium while such a forum was being conducted.

The farmers were suspicious of Stulberg and the science court idea. Virgil Fuchs recalls, "He gave the impression a whole lot of things could be considered and yet he says, but nothing could be considered. I mean he was just out here to keep the people fooled. Just to spend some time so that they could wear the people down and people knew it right away." According to John Tripp: "I felt he was hired more or less to see if he couldn't possibly get this line negotiated so that the farmers would be willing to accept it."

After several days of abortive discussions Stulberg left, promising to return after the farmers had more time to think about the science court. At that point, they were still hoping to win in the courts. The science court would surface again, but not for several months.

When the supreme court ruled in their favor, the power companies resumed work on the line. As surveying and construction crews began working their way across western Minnesota, the farmers improvised a response.

They used the tools of their trade. With their tractors, they piled boulders around the holes for the tower bases so that concrete could not be poured. Vernon Ehlers, an organic farmer from Traverse County, parked his truck with the key broken off in the ignition to block several cement mixers from getting to the tower sites. A Stearns County farmer, Math Woida, drove upwind of a utility crew and then switched on his manure spreader. Other farmers mounted their horses and chased surveyors off their land. Survey stakes would regularly disappear overnight.

The companies responded through the legal system. They tried to frighten the farmers off by filing enormous lawsuits. The company lawyers drew up half-million-dollar civil damage suits against five farmers, Scott Jenks, Math Woida, Vernon and Darus Ehlers, and Dale Tobolt. Then, at their request, district judges

Thomas Stahler and Paul Hoffman issued temporary restraining orders, enjoining "all persons" from physically obstructing, delaying, harassing, hindering, or otherwise interfering with the construction activities. Violations would bring a contempt of court citation or a misdemeanor charge, punishable by a maximum of ninety days in jail or a five-hundred-dollar fine or both.

The lawyers arranged to have their own suits and the courts' restraining orders delivered together by county sheriffs to each of the five farmers in one official-looking package. The suits and court orders indicated they applied to the five individuals and "unnamed others."

The lawsuits had a chilling effect on the protest. Five hundred thousand dollars was a lot of money, enough to wipe any farmer out. Even if the odds were strong that the companies would lose such a suit, the farmers could not be sure. They certainly had not had much luck in the courts. Jim Nelson recalls, "That had the effect of stopping everything in its tracks . . . people figured that the courts were always leaning the way of the power companies, so even in a suit like this, just from our experience with judges and the courts, we thought the odds were better that the courts would go with the power company than us."

Some of the farmers, including Scott Jenks, dropped completely out of the protest. Dale Massman, Pope County farmer and neighbor of Jenks, reflected on the suits: "I think those damage suits were just for a scare. The scare worked pretty good against Jenks. You know he's about near retirement age and the lawsuit would break him completely, and he's not young enough to start over. I think that's what probably scared all the farmers, they would have been working all their life for nothing."

Even two years after the damage suits had been filed, Scott Jenks and his wife Lorraine were reluctant to be interviewed.

Lorraine: "Our lawyer told us that he's not supposed to tell anything. We are not supposed to talk. He's absolutely restricted. And that's why I didn't want him to talk today."

Scott: "Well, you see, there is this five hundred thousand dollar lawsuit against me. I couldn't go to any meeting where there was deciding or make any decisions or give any comment."

Lorraine: "I'll never forget it. On Tuesday we were to the funeral for my aunt down in Iowa and we were driving home on

35W at 4:00 P.M. in the afternoon, we were just leaving Albert Lea and the funeral was just across the line in Iowa. And on the four o'clock news we heard about a lawsuit against the others and Scott Jenks. Well, can't you imagine how you felt hearing that in the car—five hundred thousand dollars each. Well, what would you say? No way would I have ever gone to my aunt's funeral if I had known about this. You don't know what it's done to you. You haven't felt like the same person since. You could stand the tower out there, if you didn't know they had done that injustice to you."

Scott: "The reason they did this was because they thought I was the number one man. But I wasn't. They thought if they could get me out of the picture everything would just scatter out and they would have no trouble. . . . To go to a Christian church where there are real believers and this gets throwed at you and they know it from California to Pennsylvania, they know it all over. That really hurts."

Lorraine: "We didn't even feel like going to church after getting that thing on your head."

Scott: "They just wanted to get me out. And they did it. It gives you an inferiority complex. You don't feel like doing anything, even in the community any more. And it's just not our way of life to be in any trouble."

Scott Jenks, who led the farmers at Constitution Hill, never did participate again in the protest. His health declined and he spent the better part of a year in the hospital. In the winter of 1979, he was killed in a tragic accident while felling trees on his farm.

The damage suits also had an impact on the "unnamed others" among the farmers. Witness Jim Nelson's reaction: "I supported civil disobedience kinds of activity, but about the time they started handing out these half-million dollar lawsuits, I was concerned that they'd hand them out to some people in Grant County. I supported the people that did it, but I was concerned that I might be responsible for leading someone into a situation where they'd get one of these lawsuits and I didn't want to be responsible for that."

On November 8, the farmers received another setback when the Minnesota Department of Health issued its long-awaited report on the powerline. REPORT: POWER-LINE HAZARDS UNPROVEN was the headline in the *Minneapolis Tribune.* The opening paragraph of the article read, "There is not enough evidence to show that power lines are harmful to health, the Minnesota Department of Health concluded Tuesday." State agency and power company officials all announced that they were gratified that the evidence showed what they had been contending all along—the line would be safe.

"Powerline hazards unproven" was the message that was communicated to the general public. However, an equally valid alternative way of reporting the conclusions of the study might have been POWERLINE SAFETY UNCERTAIN. Robert S. Banks, who headed the health department study, has summarized its conclusions this way: "We concluded that the available evidence was not sufficient to indict transmission lines as a hazard; however, much of the research was of poor experimental quality. It was not directly applicable to transmission line environment, and it was not—much of it was not—sufficiently encompassing to give us a high degree of confidence that the lines are safe."

In an interview at his University of Minnesota office, Banks elaborated on what the study had found: "I couldn't find any negative health effects, but I would not give the line a clean bill of health. . . . There is a question about electric shock and long-term biological effects from breathing air ions. I don't have a very comfortable feeling down here (pointing to his stomach) about these questions."

In that same interview in August 1978, Banks explained that he was keenly disappointed about the way his study had been reported and used: "In government, form is more important than substance, which is one reason I am no longer with government." He felt that more research was needed, since almost all the scientific literature on health and safety effects of transmission lines dealt with AC not DC lines: "We owe it to the people out there to get this out on the table, rather than whitewash it. We owe them some substance on this."

One reason for the public confusion about the health department study was that its report was not as clear and direct as it might have been. In our interview with Banks, we asked if he would summarize the principal findings of his study. In particular, we suggested that he make a chart on the blackboard, laying out the possible hazards of DC transmission lines, annotating each with a symbol indicating whether his study concluded it was *definitely a problem* (P); *possibly a problem, further research is needed* (?); and *definitely not a problem* (NP). He was able to respond immediately with the following chart:

electrostatic		space charge	
biological	shock	biological	shock
long term exposure to electric fields	steady state transient	air ion inhalation	steady state
?	NP ?	?	NP

There were no hazards which Banks identified as definitely a problem, but he specified three areas of uncertainty: biological effects of long-term exposure to low-level electric fields, shocks due to transients (sudden changes in the line current), and breathing air ions. The latter two effects he felt could and should be the objectives of immediate investigation. He personally was concerned with investigating air ion inhalation.

On another occasion, when interviewed by a lawyer representing some of the farmers, Banks stated, "our major concern in that report—with respect to direct current transmission lines—was in the area of air ions. Unlike alternating current transmission lines, direct current transmission lines produce air ions which can migrate away from the conductors. Air ions have been a controversial subject in the biological and public health literature for some time, and questions concerning them have not been resolved satisfactorily . . . the point I want to make is that it remains an open question, and we feel that—or, correction, felt at the time of this report—and I personally still feel that we need more information on the air ion concentrations, densities, species present, and so forth, as well as further laboratory work to assess that entire question."

In the health department report, there was no such chart summarizing the conclusions and there was no such direct statement of "the major concern." Within this lengthy volume are the concerns Banks expressed, but it would not be easy for a non-expert to read the document and pick them out.

Banks hinted at the problem of clearly stating, in an officially sanctioned study, the possibility of hazards of a project as controversial and politically charged as the CU powerline: "In government, your options are limited." There were constraints: the politics of the governor's office, the problem of going outside an agency's mandate—"if you go outside it, you are in deep trouble" —and Ronnie Brooks—"she was in the driver's seat." The government, Banks believes, "maintains the status quo, an orderly process; there is the public and there is the private sector and the government—both are your enemies."

The farmers would have agreed. Five hundred farmers, most of them from western Minnesota, held a rally on November 16 at the state capitol in St. Paul to protest their treatment at the hands of the courts and state agencies and to demand a moratorium on the CU line and a proposed NSP power plant in southern Minnesota. The rally was sponsored by an organization called CO-REG (Coalition of Rural Environmental Groups), led by farmer-physicist Wendell Bradley, a professor at Gustavus Adolphus College in southern Minnesota.

Few of the protesting farmers expected immediate results. They were seeking visibility and public understanding of the way they were being treated by their government. Holding up signs that proclaimed OUR COURT SYSTEM IS DEAD, they attacked the recent court injunctions and civil damage suits as "official lawlessness," and appealed to other Minnesotans for support.

At a press conference in the capitol, some of the protest leaders, out of desperation, called for a science court to study possible health hazards of the line, moratorium or no moratorium. They met with Governor Perpich, who was anxious to set up a meeting with them to pursue the proposal.

The rally in St. Paul did not answer the most critical question confronting the farmers—how to keep the protest going in the face of the court injunctions and civil damage suits. With the prospect of a $500,000 lawsuit, most of the farmers were uneasy about any kind of direct protest activity. They had to decide what to do next. It was a turning point in the struggle. The answer came from an unlikely quarter.

The media called George Crocker the "bearded urban radical." Crocker was well known in Minnesota for his antiwar activities during the Vietnam war. He himself had spent eighteen months in federal prison for draft resistance. The twenty-eight-year-old Crocker and a small group of "Cities protesters" were to play a critical role in the farmers' struggle. Crocker recalls: "It was in the early part of 1977 that I was first involved. I knew there was a big crisis coming in energy in the country. 'What is anybody doing?' I see the farmers are fighting right here. I watch the papers and follow it. I'd go to the capitol and start talking with people, and ask how urban people could provide support. I was very consistent showing up. Sometimes I would take the bus back with people and spend a day on their farms getting to know them better. The key thing was material aid. Anything I could write for them or look up for them I would try to do."

Crocker saw that the half-million-dollar lawsuits threatened to stop the protest. "An alternative was just to roll over and die, which was what a lot of people were thinking was what they'd have to do—'We've looked at what we've been through, the courts have decided against us, the governor says "sorry, no deal," the legislature passes a bill that doesn't apply to this line, and the utilities are coming down with their lawsuits—if you try to do anything, they'll come and take away your farm.' " He felt he could offer another way.

Dick Hanson, a young farmer from Pope County, remembers the first time George Crocker was presented to the farmers: "Well, we had a big meeting and I remember he wore his red bandana, stuff like this, and long hair and beard, and obviously there were a lot of conservative people who were still down on antiwar protesters and thought we should have won the Vietnam war, rather than pulled out. Anyway, Alice [Tripp] had introduced him —that he was here and was willing to offer the nonviolent approach and some of the experience he had. I remember one farmer, after George Crocker said he was absolutely against guns and violence. He got up and said, 'Well, I got my double-barreled shotgun and I'm gonna stand there and they step one foot on my land they're going to get blasted.' And then, since he was from Minneapolis, there was a woman that got up and says, 'How much land do you have? Is there any powerline crossing your land? What business is it of yours sticking your nose in here?' At that point Alice Tripp spoke up and really kind of put the crowd straight

—that we got people coming out here willing to help and that we'd better not start attacking them. And it was really quite good. Of course, that suspicion of Crocker stayed a long time; probably there still is some."

Crocker pointed the way by exemplary action. He and five other powerline protesters from the Twin Cities were arrested November 22 near Elrosa for disrupting surveying work in violation of a court order. As the deputies attempted to read the terms of the injunction before citing them, the six young men and many of the farmers who had gathered on the gravel road by Alice and John Tripp's farm drowned them out with verses of "This Land is Your Land." Crocker was satisfied that this action showed another way: "The other way was to stand up in the face of that court order. How many people are they gonna sue for five hundred thousand dollars? Just to go through that action and say, 'Okay, go ahead, file your suits. How are you gonna take five hundred thousand dollars out from someone who's working for wages, who lives out in the country, or how are you gonna take five hundred thousand dollars off a bunch of hippies from the Cities, how you gonna do that?' It just takes the wind right out of their sails almost as effectively as they took the wind out of the sails of those farmers by filing those damage suits."

Jane Fuchs thought the nonviolence workshops which George Crocker and his friends from the Movement for a New Society conducted were "weird." But she credits the "Stearns County Six" with renewing the protest: "That kind of started the whole thing out again. Well, it kind of brought out that people can stand up for their rights."

Pope County again became the center of the protest. At meetings in the Lowry town hall, farmers spoke against the line—to other farmers who needed little convincing—and planned their days' activities. The group had no formal officers. Steffen Pederson explained: "When they tossed those lawsuits against five farmers, we quit having a leader because we didn't want anybody to be pointed out. So whoever wanted to, and there were plenty of them, could always call a meeting to order and when he was done saying what he thought, someone else would want a say and it went real smooth. Some of us weren't very good talkers; others of us could talk real well. You know we don't all have the gift of gab."

Alice Tripp remembers those days at Lowry: "That's what is important about Pope County—people speaking for themselves. As one lady said, 'It's nice not to have people lecture at you, everyone can stand up and speak for themselves.' People get so caught up in the issues they forget about themselves and are no longer self-conscious."

The gatherings at Lowry grew in size and intensity as the farmers debated what to do. Many still feared the five hundred thousand dollar lawsuits. That feeling changed dramatically on December 15.

On that day, three farmers, John Tripp, fifty-eight, Randy Fischer, twenty-one, and Steffen Pederson, sixty-four, were arrested for civil disobedience. John Tripp has described what happened: "The workmen were welding and bolting towers together and we went into the fields where they were working. The private security men came over and called the sheriff. The sheriff came and asked us to leave. We had gone out from Lowry hall to put on a demonstration where they were building towers. We planned the arrest right then and there, kind of on the spot." Alice Tripp elaborated: "John got arrested and then Randy Fischer stepped up and got arrested—he wouldn't leave. I wouldn't leave either, but he wouldn't arrest me. Then Steffen Pederson stepped up and said, 'Can I have one of those?' "

Pederson explained: "I saw them [the deputies] writing a ticket to two guys I really didn't know. But I knew they weren't from Pope County. And I thought if we could get this into court, we would get some publicity. And we should have someone from Pope County, I thought. So I said to the deputy, and I've known him for a long time—him and his dad were raised out here—'What kind of tickets are you writing?' And he said 'Tickets for trespassing.' And I said, 'Could I be getting one of those?' And he says, 'Yeah, you could.' And I says, 'I'll take one.' "

John Tripp doesn't know exactly why he was willing to get arrested—the first violation of the law, "not excluding traffic tickets," in his life: "To resist authority I guess. Just to show our displeasure with what was going on. That was all. It was no real strategy. I don't know why. I'm kind of hard pushed to answer. Really, we never thought we'd start a movement or anything."

But they did start a movement. In the coming winter, many farmers would resort to disruption and civil disobedience to stop the powerline. John Tripp, Randy Fischer, and Steffen Pederson, with or without a conscious plan, changed the course of the protest. They were the first farmers to be arrested since the civil damage suits. That they were willing to take that step in the face of a half-million-dollar lawsuit inspired the others.

More than 250 farmers converged on Lowry Saturday, December 17, two days after the arrests, for the biggest rally since Constitution Hill. The three who had been arrested were greeted with rounds of cheers and handshakes. The farmers vowed that day to step up their protest in the fields and demanded that Governor Perpich set up a science court by January 1, 1978.

The science court possibility had remained alive throughout the fall months. Governor Perpich continued to push it. He was quick to follow up on a statement by protest leader Gloria Woida at the November 16 rally in the capitol that she favored a science court, even without a moratorium on construction.

Perpich responded by personally selecting a group of the more moderate protest leaders and inviting them to meet with him on November 19 at the governor's mansion. There he appealed to them to participate in a science court on health and safety issues without insisting on a moratorium. The group agreed and newspapers and televisions across the state proclaimed that the science court was "on." All that remained was to get the power companies to agree and, given the public relations climate created by the governor and with the full weight of his office behind the proposal, it was almost a foregone conclusion that they would have to go along.

On November 30, with great fanfare, the governor invited CPA and UPA officials to the mansion to meet with him for dinner and discuss the science court. It was like having dinner in a goldfish bowl. Reporters, photographers, and TV crews, all of whom were treated to dinner, hovered around the dinner table, which was outfitted with microphones to amplify table conversation through loud speakers.

Ronnie Brooks explained the procedure and value of the science court to the company officials. They responded that state agencies and the courts had already considered all the issues. They questioned the need for an "impartial" science court by asking Brooks, "Why would a state agency not be impartial?" She responded: "I would not say you should believe all state agencies all the time." Ted Lennick, manager of CPA, stated, "We've done what the state has required of us over and over again and now you are asking us to do more. They don't care one iota about anything unless it says 'no powerline.'" Governor Perpich warned Lennick, "If you don't find a way to objectively answer these questions, they will keep coming back." He urged the companies to take part in a science court.

The co-op officials asked whether they could leave the dinner table for a "caucus." They returned fifteen minutes later with xeroxed copies for the press of a seven-page statement they had prepared before the meeting, rejecting the science court. Much to everyone's surprise and the governor's chagrin, the companies turned him down, saying, in effect, "We have already won in the state agencies and won in the courts; why *should* we participate in this *ad hoc* proceeding?"

The companies' rejection brought them much negative press. If there were no health problems as they stated, why were they unwilling to be part of a science court? Handing out a previously prepared statement, after pretending to consider seriously the governor's proposal, was a public relations disaster. Evidently company officials realized they had made a mistake. Four days later, Perpich held a private meeting with them and announced they had conditionally agreed to a science court provided it were restricted to health and safety issues, provided there would be no moratorium on construction, and provided "no additional restrictions will be placed on the operation of the line should the science court fail to complete its deliberations." Although construction would continue, Perpich urged that the protests stop. He issued a statement, saying, "Most importantly, I urge those individuals who are obstructing the construction of the line to cease all that sort of activity while the science court is in progress."

To the governor and the power companies, these conditions seemed reasonable to insure that the science court not be used by the farmers as a means of delaying the line. To the farmers, however, they seemed to provide an incentive for the companies to delay in the science court proceedings, using them to detour the farmers' efforts away from protest in the fields, while going on to complete the line. Furthermore, many of the farmers interpreted the companies' acceptance of the science court as evidence that their obstruction tactics were working.

Finally they were angry with the governor for deciding himself which farmers would meet with him the week before, rather than allowing them to select their own representatives. Many had been upset when the governor's hand-picked group had publicly agreed to a science court without a moratorium.

At a large, emotional gathering in the Lowry town hall the next day, the farmers vehemently rejected the companies' conditions, now stating they wanted a science court, but only if construction of the line were halted during the proceedings. Steffen Pederson reflected the prevailing sentiment with a warning: "If necessary, we will hold a science court out in the fields."

As the New Year approached, Rudy Perpich had a great deal to think about. He had spent more time on the powerline during his first year as governor than on any other item of state business. In spite of his personal commitment to finding a solution, the situation had become even more explosive.

Jim Nelson has suggested why Perpich failed: "I think that initially he had no idea that it was that hard a problem. He thought that if you went out and talked nice to the farmers that we could be talked into the powerline. A lot of people will still say, 'Isn't it a fact that it's not the powerline being so bad, it's the way the powerline people approached you. They treated you badly; if they just treated you nicer, everything would have been okay.' Of course, it wouldn't have, as long as they still insisted on building the powerline, which they did. That was, I think, Perpich's idea: if he came out and talked nice to us, then pretty soon we'd say, 'Well, it's not so bad after all.' I think he just misjudged the whole situation."

The power companies had stopped surveying and construction for the holidays, but when they returned in January, Rudy Perpich knew the farmers would stop them. One year before he had declined to call in the National Guard or state patrol, but now he

had run out of options. He was faced with a choice of declaring a
moratorium on the line, which would surely be challenged by the
companies in the courts, or calling out the guard or state troopers,
which would surely be challenged by the farmers in the fields.

John and Alice Tripp
Profile

John and Alice Tripp live on a 200-acre farm southeast of the town of Elrosa. They have just retired from active farming. But every time they look out their kitchen window, they see the powerline in the field back of their house. People like the Tripps were the backbone of the protest. Their openness and integrity and determination attracted many others to their cause.

Can you tell me a little bit about your background?

Alice: We both grew up in the Iron Range [northern Minnesota mining country]. It was a wonderful place to grow up. There were all kinds of ethnic groups and the towns were too small for groups to separate from one another. I am a Finn and married John, who is Irish.

John: I grew up in Hibbing. First, I went up there two years to junior college, and then I went to the University of Michigan. Then I got a job in Detroit. I worked for an engineering company called UOP—it helped build refineries. Then I went to work for the government. We lived in Colorado and then Washington, D.C.

Alice: John, tell him what you worked on. It was pretty interesting.

John: I worked for the Bureau of Mines in the Department of Interior, evolving methods of extracting oil from oil shale in Colorado. The oil industry fought us as hard as they could and when Eisenhower was elected president, before he took office, there was a picture of him playing golf with Bill Holiday, who was president of Gulf Oil Corporation. When we saw that picture, we knew our program was over with. Well, I got sick at the same time—infectious hepatitis, which sort of knocked the stuffings out of me—and then we decided to go on the farm.

Why did you want a farm?

John: I always did, and I can't really describe it. I probably would have bought one in Colorado, but we didn't have enough money.

Alice: Why did you go into chemical engineering, John? He was good in chemistry. But tell him, John, that you were nominated as one of the chamber of commerce's ten outstanding young men in government when you were in Washington, D.C.

But you left all of this to almost start from scratch in farming?

John: You're right.
Alice: You bet, you bet.

How did you end up farming here?

John: Well, I finally thought I had a farm bought in Michigan —in the western part. I telephoned the night before and the fellow said, "John, my neighbor who I went to school with wants to buy the farm and I have to sell it to him." So then we decided the heck with it and we would come back to Minnesota. I quit the job and came back to Minnesota.

Alice: What he did was he got soil maps and he said okay we're going to live where there is good soil. This is a good soil pocket. We also looked for places with lots of lakes.

John: Well, we bought this farm in 1957—the equipment, animals, and everything. It was a dairy farm. I had never been on a farm before, so I really didn't care, except I knew I wanted animals.

Alice: The only time he saw a cow milked was in Colorado. A lady tied a cow to a tree and milked it. That is the only cow John ever saw milked. So he read these books. He came to the farmer and he said, "Do you have manuals for that equipment out there?" He had to see that manual to see what the equipment was for, he didn't dare tell the owner that. So he would read those manuals and figure out what machinery was for. And he used to walk all around this farm with a clipboard making notes and learning. It was a pretty risky thing.

How did you feel about the move?

Alice: Well, we were scared, but he was stubborn. I knew after fifteen years that is what he wanted. I didn't want to live on a farm. I never had lived on a farm either. But it was fun. It was wonderful. It has been worth it. I think the kids feel that way, too—they are very attached to the farm.

John: I have no regrets. To me I wish I started farming a long time before I did. The outdoors, I enjoy the beautiful days and the freedom. Not being on a rigid schedule, being able to work hard, but go to town or go to the lake when you want.

If you were in your thirties and farming, would you sell the farm now?

John: That's a hard question to answer. The line would be a constant reminder not so much of defeat, of not being able to stop the line, but of all the things that went into it. The way the government handled it and the way the power companies handled it and the little regard they had for people. I think it would be sort of hard to live here.

Alice: And yet how can they move? A neighbor said to me, "Where are we going to move to? There is no better land than what we have here." We happen to have a pocket of really superior soil here.

Is that the root of the protest—this feeling about the land?

John: There is a feeling toward land not generally appreciated by outsiders. They [the farmers] think everything of the land. They just feel so protective of the land. And most of them got it from their fathers—so they sort of feel it's their job to improve it and turn it over to one of their sons and so forth. With this power-line they see this won't be possible. It won't be an acceptable place to work and live. More lines will come. What we're seeing here is the use of troops to get this powerline built. This is in the final analysis an out-and-out attempt to break spirit of the people, and once done, this will be where they will run the lines. They're not going to fight the fight in a new place. They're going to come right through here. And it makes me feel so bad because they will probably succeed. The protest has been hard on farmers. The meetings until 2:00 A.M. and up at 5:00 to do chores, the constant meetings day and night—it has been awful to go through. I don't know how long people will have the persistence and will to fight back.

Many people say the powerline is necessary to meet our energy needs and that you are standing in the way of progress.

John: There has been a very great shift in people's attitudes toward progress. People are a lot more concerned now about their air and their water and farmers about their land. I think it is the power companies and the government that says we need the powerline and most people are way ahead of their government. For example, I just read the other day in the latest issue of *The Farmer* magazine an article from somebody from the university and this fella said sugar beets could be mechanized completely, with lots of herbicides, all so you wouldn't have to use labor. And this article was written with glee—you wouldn't have to use labor, this would be real progress—big machinery, big investments, that is the way to go.

I think the day of these articles is drawing to an end. I think people are questioning it. What is going to happen to all those people who do that work? Is it so wise to put people out of work? Is that real progress to use a lot of herbicides? I think people are saying, "What effect is this progress going to have on our land? What effect is it going to have on our water?" I think people are really asking questions now. And to build a big powerline in the name of progress I think is old-fashioned.

Alice: And I think farmers said that at the very beginning. We don't build a barn any bigger than we need to.

In looking at the protest in retrospect what would you have done differently? What would you have changed?

John: I would have brought in at least another lawyer. We should have had more legal help. We had a really good case legally —but we had the problem of never having anything to do with lawyers before, and finances, too. I wish we would have had this Charles Dayton [a well-known environmental lawyer] three or four years ago. I still think we have good legal arguments.

Does this mean you regret the civil disobedience?

John: Oh, no, no, no. I think the civil disobedience is an integral part of the protest, a necessary part of it. I don't regret that at all. I think we have been correct about that—correct morally and legally. We have every right to resist.

Do you think you have had any effect, or has the protest so far been a failure?

John: Even if we do lose and get the line I think that we have done a tremendous amount of good for farmers in general. So many farmers have told me that, too—that they think it is absolutely remarkable how well the farmers have stuck together. Most farmers say it has never happened before in their lifetimes —I mean the older farmers. I think we have showed farmers we can fight these things.

Alice: I think people have changed, don't you, John? Don't you think there has been a big shake-up with people like us who didn't really think much about the government, about their own lives, because they kind of had their own way? With farming you had a lot of independence.

Do you think a science court might have helped resolve this conflict?

John: No. I felt Mr. Stulberg was hired more or less to see if he couldn't possibly get this line negotiated so that the farmers would be willing to accept it.

What about Governor Perpich?

Alice: Perpich and Stulberg both said, "The first thing we'll take up is need." They haven't repeated that.

Why do you think the governor proposed a science court?

John: I don't think he was sincere about it at all. The only thing he could see in the science court is that it would sort of mollify farmers or at least sort of defuse them—because they would think the line's being built, but on the other hand if it is unsafe it will be taken down or something.

Then the health and safety effects were never proven and that was the weak side as far as he was concerned. Having a science court would convince people throughout the state it was being done correctly. That's how I think.

Alice: Perpich keeps saying publicly the only unsettled issues are health and safety—so he stuck us with that. We didn't talk just about health and safety. That's just another media mix-up.

Alex Olson said, "That is just a handle you guys have—when you talk about health and safety." Well, in a way it is—it is a handle—but not just a handle. It seems like all the other concerns we had didn't even count.

Do you regret the protest?

John: I wouldn't have done it any other way. I wish we could have found a better way to do it so we could have been more successful. I can think of something. When we had our meetings, we should have gotten more people involved.

What difference would it have made?

John: I think if people knew what the issues were, people would have supported us more.

Alice: Go back a little bit. Could we have done something even earlier on? Before the construction permit? Why didn't anybody go down and occupy the capitol a long time ago?

John: Because I don't think farmers are that type of person, Allie.

Alice: The heck we aren't, John. Do you know what the Whiskey Rebellion was? That was back in the time of the Revolution. Those were farmers that marched and disrupted.

What would it have accomplished if people had gone down and occupied the capitol?

Alice: To go back to John's idea, we should have had a lawyer. Well, somebody should have been here to tell us that we had to fight harder before June 3, 1976, because that was the point where they granted the construction permit. We should have gone down to the state capitol and made it difficult to grant that permit. The hearings were not enough. All that was a set forum—here are the listeners, the board, here are the people speaking their piece, and then we'll tell 'em to go back to the cornfield and be quiet—and that's exactly what they did—tell the public, "We've heard the farmers and have had lots of hearings." Somehow or another we should have rocked the boat so much they couldn't have granted the permits. Of course, that is hindsight again.

Or maybe if we had more confrontations before the supreme court hearing, if the protest was bigger then, maybe the supreme court decisions would have been different?

John: It would have been hard, because people depend on the courts.

You don't think there were any other channels open to you?

Alice: Well, if you look back, I don't think there were.

What do you think is going to happen to the protest?

John: I'm kind of worried about it. I personally can't think of anything to do. I'd like to keep protesting and slow down the construction work until elections (1978). I just can't help but think we might make a political issue out of it, and that there is still a chance we can win somehow if we keep going.

Alice: Do you think there is any value in keeping the protest going? And what is the cause?

John: Well, it is a lot more than the powerline, isn't it? It is the fact the farmers feel they are not getting an economic break. There is a tremendous amount of dissatisfaction with the government.

Alice: John, one thing that bothers me is that you don't give the companies enough credit for the power they have. You blame the government, but refuse to acknowledge how powerful the utility companies are.

John: To me a company is organized for one purpose, that is to make a profit, and they're going to do that. I don't think you can blame people for doing what they're supposed to do. The government is where I look for something to protect the people.

Alice: You can't expect to go out and make a profit at any cost regardless of—

John: No, I expect the government to step in and stop them.

Alice: Well, you have to expect the company to have a little responsibility—that's what happened to this country—"What's good for General Motors is good for the country"—the profit motive.

John: But the governments do their bidding. The government is not performing its function which is to protect people against excessive abuse.

Alice: You think NSP isn't as bad, but they are; they've done just as gross things, and I think it's utilities all across the country that are trampling on people.

What can people do about this?

Alice: A lot of people keep denigrating what happened at Lowry —all negative, and that's not true. I insist on saying that we did achieve something, and maybe it's the start of bigger things. We have to do a lot of overturning, a lot of turning upside down; meanwhile the companies are moving in as fast and hard as they can—dictating our energy policy, getting natural gas and oil deregulated, and that sort of thing. One thing you have to do is make people aware of things that are happening. Even though it's not on your land, it's happening to you. And that's a hard thing to do. It turns out that when the line runs on your land, you suddenly become a revolutionary because you realize what's happening that you didn't think about before and it's happening to you.

Some people are afraid of what their kids are learning—that they are learning to defy the law. I suppose that is a concern of all parents. I think it is important to learn that you can and should defy the law, because I grew up with the idea you must obey. I'm not so sure that is a proper way to grow up. We were kind of accepting of everything, kind of retiring from things. I was not always willing to take up arms. I was not always so determined the world should be turned upside down.

Rudy, Rudy, I've been thinking,
Pull the bulls out of our fields—
Rudy, Rudy, graying Rudy,
I don't think that we will yield.

—from "Henrietta's Song"
by Henrietta McCrory

6   Confrontation on the Prairie

A confrontation was inevitable. When the surveyors returned to the fields near Lowry in early January 1978, they were met by a large crowd of protesters. The Pope County sheriff's department, backed by state highway patrolmen, was also on hand. On Tuesday, January 3, eight farmers were arrested. One of them, a fifty-nine-year-old woman, refused to cooperate. She was picked up bodily and carried across the field to a waiting squad car. "An old lady treated like an animal," was the way angry farmers described what happened.

The woman was Alice Tripp. She had intended to be arrested: "We went out to obstruct surveyors. They told me to move and I wouldn't, so Hebner, the assistant deputy sheriff, arrested me. There were also some highway patrolmen there. And they carried me to the car because I was told to make them carry me. It sounds kind of funny. Everybody tells you, 'Lie down—it will take four men to carry you and it will occupy more people." John said he saw me from across the field with just my head bobbing along and was quite alarmed. . . . It was a symbolic thing. We hoped eventually to get a whole crowd arrested. . . ."

The next day, a hundred farmers, stirred by the eight arrests, assembled at the Lowry town hall and went out into the fields after the surveyors. They chased powerline crews from three separate sites and then, carrying American flags, marched en masse toward state troopers guarding the entrance to a powerline materials yard in Glenwood. Scuffles broke out in the snow as the troopers tried to arrest those at the head of the march. The farmers linked arms to prevent arrests. They vowed the police would not drag any of them away as they had Mrs. Tripp.

After the confrontation the state commissioner of public safety, Edward G. Novak, flew to Glenwood to confer with Sheriff
Emmons and county attorney C. David Nelson. Emmons had requested state patrol assistance and there were ten patrol cars in
the area, but local deputies and a few patrolmen could no longer
handle the protest. The pressure was on Governor Perpich. The
entire state of Minnesota was watching through the media. Public
opinion was on the side of the farmers and they knew it. They
wanted Perpich to stop the line. John Tripp recalls, "A moratorium was our big hope."

On Friday, Governor Perpich did act, but it was not as the farmers had hoped. He delivered a brief, blunt message on KCMT, the
Alexandria television station that reaches homes throughout
western Minnesota. He said he had spent more time on the
powerline dispute than anything else; that his administration had
"gone the extra mile" to try to solve the controversy. Then he announced the largest mobilization of state troopers in Minnesota
history—215 of the state's 504 highway patrolmen were dis-
patched to Pope County.

This set the stage for a decisive moment in the protest. On
Monday the state troopers would be there in force. National
media attention was focused on western Minnesota. Dozens of
reporters and all the national television networks converged on
Lowry that weekend. A violent confrontation was a definite possibility. No one knew for sure what the farmers would do.

Monday, January 9, was a bitter subzero day. After meeting in
Lowry, the farmers drove to a tower assembling yard four miles
west of town, where they had chased away work crews the week
before. Two hundred protesters, as usual led by men and women
carrying American flags, marched toward the yard. Scores of
troopers blocked their path. A state patrol helicopter and plane
circled over the construction site. It was an unbelievable scene in
rural America.

The farmers marched resolutely up to the troopers, then
stopped, and offered each of them a plastic carnation, with five
roses for the ranking officer. The troopers were not sure what to
do, but after some hesitation, most of them ended up holding a
flower. In addition to flowers, the farmers also offered them hot
coffee and home-baked cookies. There was no violence. The

event received widespread state coverage, all very positive. But the national media lost interest and recalled their reporters and cameramen.

This was a turning point in the protest. Over the weekend at Lowry town hall the farmers had debated what to do—whether or not to confront the troopers and take mass arrests. It was possible that a dramatic confrontation, with national media present, would create nation-wide sympathy for the farmers' plight and focus irresistible political pressure on the governor and the legislature. The farmers could not stop the line as long as state government continued to support the power companies.

In the end, a steering committee, headed by George Crocker, decided on the strategy, which was to avoid violent confrontation and mass arrests. Crocker explains how they viewed the likely impact on public opinion: "Our thinking was, don't set yourself up for being terrorists—law and order is a very strong thing. We gave them plastic flowers to show we were not attacking human beings."

Another strong consideration was the committee's feeling that many of the farmers had not yet reached the point where they were ready to be arrested. A strategy of mass arrests might fall flat. Gloria Woida remembers: "I felt all along a strategy had to be used where we wouldn't turn the people off and they would have to be seasoned to the idea of confronting the highway patrolmen and getting into peaceful arrest and having them accept that. I felt that was a good way to start it off."

Other farmers were disappointed. Nina Rutledge was one: "I wanted to do something. I didn't want to stand around and give them cookies and flowers. I remember feeling terribly frustrated." John Tripp felt, "It looked like we had given up." Tony Bartos advocated direct confrontation: "When those cops first came out, I thought we should have chased them right out of here—more or less run over them and tell them that's it, get the hell out of our country. We had enough people. I don't know which ones it was that gave them flowers and that bullshit."

George Crocker, in retrospect, thinks the protesters should have resorted to massive disobedience, and that they missed a unique opportunity to generate national political pressure: "The next day the national media was gone. The mistake was we should have taken one more step—given them flowers and then walked right through them to the field of tower tops and stayed there."

The debate over strategy that took place before that fateful Monday would continue throughout the months the state patrol remained out in western Minnesota. Some of the protest leaders advocated civil disobedience and mass arrests. Others feared mass arrests would scare away many farmers and undercut the protest. Still others contended the protest should be militant. Passive civil disobedience was an alien idea to many of them. It was like getting arrested for doing nothing. Math Woida and Tony Bartos were among those who felt that way. According to Woida, "What the farmers didn't want to do was sit down in front of something and just be drug off. If they were gonna be arrested, they wanted to have the feeling of being arrested for really doing something."

Wednesday, January 11, was the day of the "Rutledge Eight." It was a complicated and confusing day which revealed the serious differences over strategy within the protest movement.

Dennis Rutledge was known among the farmers as a brilliant field leader. He had proven himself at Constitution Hill and was later elected president of FACT. A graduate of the Naval Academy, Rutledge had been a career Navy officer for thirteen years before coming to Lowry in 1975. He and his wife Nina were newcomers in a community where people were still regarded as "outsiders" after living there for twenty years or more. He had become active in local DFL politics and others in the community sensed that he had political ambitions. Governor Perpich had picked their farm for his first visit. Nina had subsequently been named to a committee advising the governor on potential appointees for important positions in his new administration. As a consequence, some of the farmers were suspicious of the Rutledges.

Both Dennis and Nina were strong advocates of mass arrests. Dennis has explained why: "My feeling was that a hundred people arrested would make a real difference—the media would blow it all out of proportion and we would get the publicity we wanted. I just felt we had to get the word to the rest of the people in Minnesota and we couldn't as a handful of people. With more interest there would be more pressure. The governor was more our intention—he or the legislature could put a halt to it."

That morning everyone met as usual at Lowry town hall. Nina Rutledge will never forget what happened: "We got up and went into town and there was a huge meeting. And when we got there, I knew something was going to happen that day; I just had a funny feeling. I hadn't slept all night. Sure enough, when we got there, somebody came running up and said, 'This is the day—they're on your land.' I remember I started to shake. And they announced, anybody who is willing to get arrested wear these armbands, and people were putting on armbands of different colors. So Denny got up and showed where the surveyors were and where the roads were and how you could get on the land and where the police cars were and how to get around their roadblocks and what to say.

"We got out there and there were hundreds of people and they were carrying the flag of course and they just surrounded the surveyors and of course they—the highway patrol—were shoving some people and some people were really mad. It was frightening because there was such a mass of people—hundreds of people pushing in—that these surveyors were, I am sure, scared for their lives. It was the kind of situation where you could feel almost anything could happen and nobody would realize what had happened because it would happen so fast. Denny just stood there and screamed at the highway patrol."

Dennis continued: "I was very calm and cool, I think. What I was saying was that these guys were on my land and they did not have identification. I asked for their identification. This guy says, 'I am not going to show you anything.' I said, 'Okay highway patrol, I'd like that man arrested; he is trespassing.'" Nina: "They said, 'No, we can't.' 'Well,' Denny says, 'then I am going to put him under a citizen's arrest.' Denny put him under a citizen's arrest and they didn't do anything about it, but they escorted the surveyors off. We were all happy and clapping and cheering and decided to meet back in the hall."

The survey crew said the companies would call the attorney general's office to find out whether they needed identification papers and that they would return to the Rutledge farm at

2:00 P.M. The farmers scheduled a 12:30 meeting at the town hall to plan strategy. The Rutledges planned to block the surveyors and hoped the others would join them.

That did not happen. Shortly before two o'clock, Math Woida rushed into the hall and announced that the state patrol was setting a trap out at the Rutledge farm. Woida recalls: "From Glenwood I took the back roads to Lowry and I saw ten or fifteen state patrol cars along the road around the Rutledge farm and I just figured the way everything was set up it was a trap. If everybody had been arrested then the protest would be pretty well over, because everybody would have been scared." Most of the farmers decided not to go to the Rutledge farm and instead demonstrated that afternoon at a steel storage yard in Grant County.

This well illustrates the disagreement that then existed among the protesters about mass arrests. What Math Woida viewed as a trap was what the Rutledges hoped for. The Rutledges expected the state troopers to make arrests if the farmers resorted to civil disobedience; they felt the farmers should make a stand and be arrested. Gloria Woida explains the counterview that prevailed at Lowry that day: "The highway patrol had just been out there a few days, right? And Dennis was really pushing hard to get the people out. I felt that I didn't want it justified for Governor Perpich to have his little redcoats out there and have a mass arrest a couple of days after they came—I didn't want to have Dennis Rutledge to use the people to justify Rudy Perpich's decision to have the highway patrol out there. I didn't want the people used that quickly. The farmers had never been arrested before—they'd have never come back again to Lowry."

The Rutledges were not entirely alone that day. A few protesters returned to their farm with them. Followed by Nina and the others, Dennis walked through the cow yard to the back field where the surveyors were working. The troopers saw them coming and drove their cars across the frozen field to head them off. The media people, with their cameras, were close behind. Dennis has described his feelings at that moment: "We didn't want to have any incidents with the surveyors. We wanted to have a symbolic arrest. In a way, we were defeated by what had happened earlier that day. But we felt it was extremely important that my wife and I get arrested on our land confronting the surveyors. We felt that was important personally."

It was late in the afternoon. Nina and Dennis Rutledge, ages thirty-four and thirty-five, held hands and stepped in front of the surveyors. They were arrested for "obstructing legal process." The others followed: Dick Hanson, twenty-seven, Grant Hanson, twenty-eight, Earl Hauge, thirty-seven, Mary Stackpool and Amy Lee, both in their forties, all of Glenwood, and Reed Holt, a student at St. Cloud University. The eight of them stood together in the middle of the snow-covered field and were arrested. One other farmer, Alice Tripp, was there. She recalls, "It was very dramatic, their arrest. It was upsetting." Nina Rutledge remembers just how she felt: "I started to cry. I was scared and I was happy. It was all kinds of feelings mixed up—I was really confused. I felt strong about what I did, but I still had confusing emotions—even feeling guilt, that I had done something wrong, because of the way I had been raised. They took us all to the Glenwood jail and took mugshots and fingerprints. It was very trying."

Even though they were only eight, they had an impact. Dick Hanson announced he would fast until Rudy Perpich called him. Media coverage of two young farmers arrested while standing peacefully on their own land and another young farmer fasting in jail aroused public sympathy. Even more important was the response of the man who was supposed to prosecute them.

On Friday, January 13, Pope County attorney C. David Nelson made headlines around the state. Refusing to prosecute the protesters, he resigned. Nelson spoke of the "Rutledge Eight" as upstanding citizens and fine people: "They're not criminals, and that's the point. The law is clear and they are breaking the law as far as I can see, but I don't agree with the law."

He explains his resignation this way: "Well, I tell you, I resigned for a multitude of reasons, really. One of which was the fact that I thought the farmers were getting the short end of the stick. My position as county attorney was fairly clear. The law said—there are court orders—they could not obstruct the building of this line. I really still do kind of in the back of my mind think it is relatively harmless to stand in front of a transit. I think a lot of these people did this, you know, not because they're trying to be mean, but because it meant something to them. And I guess I thought the law was wrong and if you believe the law is wrong, then you shouldn't enforce it."

C. David Nelson had been the Pope County attorney for eleven years. When he resigned, none of the other six lawyers in the county was willing to take his place. They did not want to have to be put in the position of having to prosecute the powerline protesters.

These events set off a cycle of civil disobedience and arrests. During the days, caravans of cars and trucks full of farmers cruised the wintry country roads looking for surveyors and construction crews. Confrontation followed confrontation. At night, survey stakes were removed and incipient towers disassembled.

After the troopers had been in Pope County for one week, the balance sheet read: salaries and fringe benefits—$93,028.64; overtime and fringe benefits—$48,889.14; fixed wing aircraft—$1,269.20; helicopter—$1,500.00; 215 patrolmen and 84,240 squadcar miles. Instead of stopping the civil disobedience, however, their presence had intensified the protest.

At the center of the protest was the little town of Lowry, population 257, 150 miles northwest of the Twin Cities, just off the Floyd B. Olson Memorial Highway (named for Minnesota's famous populist governor of the 1930s). During the "occupation," as the farmers called it, the area was literally teaming with highway patrolmen. Troopers in squadcars guarded the edge of town to monitor all traffic coming and going. In the middle of town, a trooper hung in effigy; a half-block away was a remembrance of former governor Wendell Anderson—a dead turkey hung from a lamppost with a "Wendy" sign tied to its leg.

The Lowry town hall, a small two-story brick building, housing the fire department on part of the ground floor, became the command post for the protest. Downstairs, the walls were papered with anti-powerline posters and photographs of farmers being arrested in the fields, and there were tables of literature, sign-up sheets, and a seemingly inexhaustible supply of homemade sandwiches, cookies, and coffee.

Upstairs, anywhere from three hundred to eight hundred protesters crowded into the auditorium each morning. Everyone was free to speak. They would describe what happened the day before, make speeches against the line, debate and argue strategy, and make plans for the rest of the day. Farmers and their families regularly came from as far as a hundred miles away. In terms of political persuasion, they were an incredibly diverse group, ranging from radical to liberal to as far right as you can get. Despite their ideological differences and differences over strategy, a deep camaraderie developed; they were united in their struggle against the cu line.

One issue kept recurring in their discussions: whether or not to have mass arrests. Alice Tripp, in particular, argued for a dramatic civil disobedience action. She advocated that as many men and women as possible be arrested and refuse bail. The jails could not hold all the protesters. Such a dramatic action might produce political reverberations around the state. People all over would ask, "Why?".

Though respected by all, she was unsuccessful in persuading the group. Others were sure mass arrests would scare people off. They feared that once the troopers arrested farmers, they could threaten them with more serious charges if they were arrested again and could thereby break the protest. More importantly, many of the farmers found it personally difficult to contemplate breaking the law.

Dean Danielson, operator of the Lowry grocery store, explained that deep feeling: "You have to realize, myself included, most of the people are very conservative out here and it took a lot for them to protest. And to say, 'Subject yourself to arrest'—that went overboard for them. Back to basics. It's not nice to get arrested. You don't break the law. When you believe that all your life and you're a fifty-year-old farmer out here you just don't change overnight."

However, there were frequent clashes between farmers and troopers in the ensuing weeks and many farmers did go to jail. On Tuesday, January 17, fourteen persons, ranging in age from sixteen to sixty-eight, were arrested in two confrontations on the Dale Massman farm. Eight of the arrests came in the morning, when protesters formed a moving circle in front of two surveyors. Singing "We Shall Not Be Moved," they linked arms and circled about the transits. Thirty troopers tried to move the circle out of the surveyors' lines of sight. In the resulting free-for-all on the frozen field, several troopers and farmers were injured.

Dean Danielson, one of those arrested, recalls: "It was rough. The patrolmen came in and one of them said, 'Next time we get you guys on the ground, you're not going to get up.' It is scary because they are the law. Getting arrested is very scary. I lost a lot of sleep over it. Some people it doesn't bother; me it bothered. It wasn't something I ever thought would happen to me."

After lunch, there were five more arrests. Gloria Woida, one of the most outspoken and articulate of the protest leaders, was arrested for the second time that day: "There were about fifty people out in the fields and they had knocked over a tripod. When I got there they were scuffling with highway patrolmen, and seeing the tripod lying there, I picked it up and smashed it. I didn't see this one patrolman who yelled, 'You're under arrest," and came running toward me. He tackled me just like a football player. I let him have it with my foot and that is where the simple assault charges come from. Then George [Crocker] came up and threw himself over me and said, 'Don't you dare hit her.' Then George was arrested. . . . They carried us off to jail."

Alice Tripp, arrested in a demonstration two weeks later, described how it felt to be behind bars. She had decided on principle to refuse bail and to remain in jail until her trial. Although she was a leading advocate at the Lowry meetings of a mass-arrest strategy, she was never absolutely certain she was doing the right thing. In a letter to us dated, "Willmar Jail—Saturday," she wrote, "I don't know whether I can last ten days. And I am afraid the whole thing may appear to be grandstanding. I thought I could stick it out, but I'm no Bunyan and I'm still not sure." As it turned out, she didn't have to stick it out. Against her will, she was released on a technicality. She believes it was because the authorities decided that holding one farmer in jail would provoke the others: "At that point, they didn't want anybody in jail, but particularly old ladies."

The protest in the fields kept building. Five hundred to a thousand people were turning out—too many for the troopers to handle. Some of the farmers felt this was their best hope—to have so many people out every day that they could handle the troopers and stop the line.

But others realized that it would be difficult to sustain the momentum. For one thing, the farmers had time during the winter, but spring planting was only a couple of months away. The massive outpouring of people into the fields, day after day on the coldest days of a Minnesota winter, had clearly demonstrated the breadth and depth of feeling in the populace about the powerline. The time seemed ripe for an appeal to the governor and the legislature.

George Crocker recalls the farmers' hope: "Well, we'd been up here for all this time, everybody knew we'd been on the news, we made front page and all the television and radio stations, people knew we were around, so we thought we maybe would have a chance of cracking something open after that show of strength. After all this momentum we thought they would listen to us."

Several thousand farmers converged on the capitol in St. Paul. Some communities in western Minnesota virtually closed down for the occasion. In Glenwood, the Pope County seat, the schools let out so that teachers and students could join the protest.

They asked the governor and the legislature to declare a moratorium on the powerline until environmental and health and safety problems were carefully studied. The sheer size of the protest tied up the capitol building. At first, the governor and legislative leaders refused to meet with them. The situation became quite tense. The message *we want a moratorium on powerlines* was carved in an antique table as several hundred protesters crowded the reception room outside the governor's office. It appeared that the farmers, several thousand strong, would sit down at the capitol and spend the night.

This prospect was headed off, however. Martin Sabo, Speaker of the House, agreed to a meeting. He defused the tension by promising that the legislature would hold hearings to look into the possibility of declaring a moratorium.

Governor Perpich's state-of-the-state address, delivered the next day, devoted considerable attention to the powerline dispute. He emphasized that, "In our democracy, there is a system of law to resolve disputes. If democracy is to endure, our respect for law must endure. . . . The parties to this dispute have had the fullest access to our entire political system." He concluded, "I still regard the science court as the best mechanism for resolving this controversy. I again call on all parties to join with us to make the science court a reality."

While not enthusiastic about the science court, the farmers were hopeful about the legislative hearings, which were scheduled to start in early February. George Crocker said: "They [the legislators] worked against us in that they threw us this candy bar. They said, 'Okay, we're going to have these hearings now,' and people are thinking, 'Fine, we made it!' That was clearly on people's minds—'Well, now the legislature is going to bail us out!' "

That did not happen. What did happen was that the huge crowds that had been coming to Lowry each day dwindled to a couple of hundred stalwarts after the demonstration at the capitol. The other farmers waited to see what the legislature would do.

The legislative hearings were just another charade. From the beginning, knowledgeable legislators knew what the farmers did not know—that nothing would come of the hearings except talk. The subject of the hearings was not a bill to impose a moratorium, but rather an "advisory bill" that called for a study of the feasibility of a moratorium. Representative Ken Nelson introduced House Advisory 60, "A Proposal to Study Feasibility of a Moratorium on a Particular High Voltage Transmission Line":

> Controversy surrounds the construction of the Minnesota segment of a 400,000 kv high voltage transmission line between Underwood, North Dakota and Delano, Minnesota by rural electric cooperatives.
>
> The appropriate committee or committees of the Minnesota legislature should, at the earliest possible time, conduct a study of the feasibility of imposing a moratorium upon construction of that particular line.
>
> Study should take into account various possible consequences of a moratorium including, but not limited to, the time required to complete a study of health impacts of the line, the financial impact of a moratorium upon the cooperatives building the line, any possible liability of the state for added costs to completion, and effect of delay upon rates charged to consumers.

Four days of hearings were held before the House committee on environment and natural resources. The first day the farmers and the power companies were each given time—one hour—to make their cases. The farmers reiterated their concerns. Attorney John Drawz, speaking for the power companies, made the headlines. He declared that the state would be liable for the eight hundred million dollars invested in the CU project if the legislature imposed a moratorium on construction.

The second day, two legislative staff members gave short reports. Legislative analyst Sam Rankin told the committee that the companies' demand projections were accurate and that no alternative sources of power were available. His one-hour report was all the committee heard on the question of need. Joel Michael, an attorney with the house research department, testified on the legality of a moratorium. He said that if serious health or safety hazards could be established, there would be legal grounds for a moratorium and took issue with Drawz' contention that the state would be held liable for the full cost of the project. The thrust of his testimony, however, emphasized the potential legal difficulties the state faced if the legislature imposed a moratorium.

The farmers who crowded the hearing room were quite discouraged by this time. They began to sense what was happening. After the second session, Virgil Fuchs voiced what many were feeling: "I have to believe in this, otherwise why would I come. But it seems like they are just getting us farmers to drive up here a long ways every other day to meet between ten and twelve to keep us off the fields." John Tripp sat stoically throughout the testimony. He was tense. The frustration showed on his face.

The third day featured more testimony by legislative staffers —this time concerning health and safety and other environmental impacts such as television and radio reception. As Michael had pointcd out in the preceding session, health and safety was the one area that might provide grounds for the legislature to declare a moratorium. The testimony was so dry and technical, however, that some committee members fell asleep. What is more, it did not touch on any of the issues—air ion inhalation, transient shocks, or long-term exposure to low-level electromagnetic fields —that Robert Banks, author of the health department study identified as the possible problems.

Banks himself was disappointed with the quality of the presentation and angry that he had not been invited to testify. As he put it rather bluntly later, "If they had gotten people who knew something, they'd have learned something. Instead, they wanted to put on a show and the committee looked like horseshit."

The evening before the last session, one member of the committee predicted privately what would happen. The next day, he confided, "would be the big day" where "the whole cynical deal would be played out." He said there never would be a vote on any moratorium. An "advisory bill," he pointed out, did not require a vote. He spelled out the following scenario: somebody on the committee would move there be a moratorium; the chairman would rule him out of order; there would be a vote on the chairman's ruling; and that would be the end of it. He concluded by saying, "Nobody at the legislature wants to go on record voting on the moratorium. Tell the farmers to blow up the towers if they want to keep the powerline out."

That never had a chance to happen. Instead, Representative Nelson announced that he was withdrawing his advisory bill. In its place he offered an amendment to provide one hundred thousand dollars for monitoring of health and safety effects of the powerline. In a six-page statement, he noted that while he would like to see a moratorium, he didn't believe it possible for economic, legal, and practical reasons. He concluded: "I would like to thank the protesting farmers for bringing these energy questions—the questions of supply and demand, conservation and alternatives, to the consciousness of the Minnesota people— something this committee has been trying to do since 1972. But I would request the farmers to not have bloodshed in the fields. I believe that the state has responded as humanely and as sensitively as possible to a bad situation. Sure, there has been human error in judgment and human failure, but nobody is going to win this battle—it's too late for that—we all have our scars. But we have all learned from it. This won't happen again if all of us in Minnesota, citizens and legislators, utilities and farmers, really work together on energy problems.

"The protesting farmers have made their point—a good point. I believe we have responded and will continue to respond, not always as they wish or we wish we could, but as we are able."

The committee voted unanimously to allocate funds for monitoring health and safety effects. One member, Arnie Carlson, who was going to run for state auditor, did speak against Nelson. He made an impassioned statement about atomic weapons testing, asbestos fibers, and Vietnam, and said, "We must change the burden of proof on the government to prove there is no harm." He didn't make any motion, however. There never was a vote on the advisory bill, much less a vote on a moratorium.

The farmers felt betrayed. Rather than a full hearing they had gotten a brush-off. They could not agree with Representative Nelson that the committee had responded to their concerns, and they were particularly bitter that he changed his position and dashed their last hopes with the legislature, without the courtesy of letting them know before the hearing. John Tripp summed it up: "That was just a frame-up from the word go to me."

Phyllis Kahn, a liberal activist legislator from Minneapolis and a member of the environment and natural resources committee, knew from the start there was "no legislative solution." Most of her colleagues took the position that too much was already invested and that this powerline would have to be accepted, mistake or no mistake. Even if the committee had passed a moratorium bill, it would have been "a useless gesture." It was clear from the beginning that such a bill had no chance in the legislature.

The farmers were not experienced at the legislative game. As Steffen Pederson would explain it, "We weren't formulated for that kind of work." [A personal note: When we met with the farmers during the hearings some of the leaders were impressed that we knew where various legislative offices were. They were trying to lobby individual legislators, but first had to find out where they could find them. Our first thought was, "It is amazing that these farmers don't even know where the legislators' offices are and yet they have had the whole state government on the run for two years." Our second thought was, "Probably the reason these farmers have had the whole state government on the run is because they don't know where the legislators' offices are."]

While the legislative hearings were going on, the protest in the fields was heating up. There were two violent clashes between farmers and the state troopers.

The first occurred on Friday, February 11, when 125 protesters converged on the Alexandria Concrete Company, which supplied concrete for tower footings. Linking arms, the protesters tried to stop the concrete trucks from leaving the yard. Sixty state troopers and five Alexandria police officers tried to make them move. As scuffles broke out all over, a truck plowed through the melee. It was a miracle no one was killed. There were three arrests —Jacqueline Thurk, thirty, of Villard, John Tripp, and George Crocker. One trooper and two of the protesters, John Tripp and George Crocker, were carried off by stretcher to the hospital.

John Tripp has described what happened that day: 'I got to Lowry that morning about 11:30. Carolyn Koudela was chairing the meeting and just as I walked in she said, 'Remember, we want this to be a peaceful demonstration.' So everybody started to leave. I left, too. I didn't even know where we were going. I asked someone where we were going. They said, 'Alexandria.' When we got there—there were a lot of people—we got in a circle. In the meantime, we could see they were preparing to leave and then all of a sudden they burst right out of there—slowly, but without stopping—right through the middle of us. I remember I got in front of a police car. It was very slippery—the roads were packed with snow and ice. And it was dangerous with those trucks and cars moving—there was an awful lot of pushing and shoving all around. It was a very physical encounter.

"Some of the highway patrolmen yelled for the truck to stop. But as it kept coming, more and more troopers rushed in. There was more pushing and shoving. Then all of a sudden I saw a person [George Crocker] lying on the ground, his head next to the wheel of a police car, with two cops on top. I went over and pushed one cop off. He said, 'You're under arrest.' Gloria [Woida] yelled, 'Run!' I yelled back, 'Where am I going to go?' They started to put handcuffs on. I said, 'Take me to a hospital; I might have a heart attack.' I really didn't want handcuffs."

The "battle of Stearns County" took place on February 13, 14, and 15. It was fought on the Woida farm. Math Woida has described what happened when the surveyors came on his fields on February 13: "I got on a big tractor with a manure loader and went out there back and forth across the fields toward the surveyors. Some college kids from St. Cloud drove the other tractor with a manure spreader and five or six people in the spreader with baseball bats. The surveyors left their tripods and run. We chased them back out of Stearns County that day.

"On the 14th they come back again early. I went out there all alone at 8 o'clock with a tractor and manure loader in front. I wanted to make a trail so we could get out of there and began pushing snow near the patrol cars. I shoved snow right on top of the cars—one trooper jumped out, cause he figured I would roll the car right over. So then thirty-five troopers come. I put the tractor in reverse and they couldn't catch me. They were just hollering, 'You are under arrest, under arrest.'

"Then these other guys jumped in the manure spreader again with baseball bats—the Wadena bunch. As the troopers charged them, I came back at the troopers with the loader and they jumped out of the way. This gave those guys a chance to jump back in the spreader. Then the troopers maced them. I called the sheriff and told him they used mace on us. Ben Herickoff went up to the captain of the patrol and said, 'Mister Captain, I have a question. Can we use mace?' The captain replied, 'You have a right to protect your property.' We cheered and said we just might do that.

"The next day they got stronger mace. We had the biggest crowd out for this—over two hundred. But only a few knew what was going to happen. We got about nine big tractors. By ten o'clock there must have been 120 extra troopers. They stopped the pickup with the anhydrous ammonia tank three times, but eventually let it go through. It was a plain white tank and they had no way of knowing what was in it. The troopers were all standing on the road—lined up bumper to bumper in eighty-six cars. All at once there was anhydrous ammonia. They were so stupid, they run with the wind and the further they run, the more it come. Some sat in their cars and figured they were fine and all at once the heaters sucked that ammonia in and zoom, I think several of them left on two wheels. About seventy-five with riot gear started coming across the fields. In the meantime, the farmers, wearing face masks, charged at them in the tractors. Them troopers were diving into the snow to avoid us.

"Them troopers were scared that day. They scattered like flies. I remember hearing them say, 'I didn't think those damn farmers were going to try to run us over with their damn tractors.' "

Media reaction to the violence in Stearns County was quite negative. It probably adversely affected public sympathy for the farmers. Throughout the protest, state-wide public opinion polls consistently favored the farmers' cause, but indicated a much more mixed reaction to their tactics. Furthermore, even many of the farmers were upset by what had happened at the Woida farm. They knew that anhydrous ammonia, commonly used on the farm, was an extremely toxic chemical. Less than fifty people showed up at the Lowry town hall the following week.

Another bad omen was the conviction of John Tripp. Three weeks after his arrest at the Alexandria cement plant, Tripp was put on trial. After two hours of deliberation, a jury of three men and three women found Tripp, aged fifty-nine, guilty of "intentionally inflicting or attempting to inflict bodily harm against another."

This was the first time a farmer had been convicted. Tripp recalls his surprise when it happened: "Everybody assured me they can't find you guilty. When we had dinner in Alexandria waiting for the jury verdict, that was our attitude. When I was convicted, I don't know, I was kind of shocked." The judge must have been a little shocked, too. In sentencing Tripp, he stated, "The jury found you guilty. I may not have arrived at the same conclusion, but the Constitution requires that I go with the jury's verdict." The sentence was ninety days in jail, with eighty-five suspended and five days stayed and a fine of five hundred dollars, three hundred of which was suspended.

Perhaps the judge was influenced by Tripp's impassioned statement just prior to the passing of the sentence. Tripp challenged the need for the line, discussed the missing transcripts, described the potential health hazards, and told of the "silence from state government" in the face of these concerns. He concluded: "Your Honor, we are told continually by construction workers, surveyors, and by the sheriff and his deputies that they are just doing their job. I submit to you that we farmers also have a job to do. This line jeopardizes our jobs by ruining our farm fields and makes our place of work unsafe. Our jobs, our homes, our health —is the law right when it stands beside this corporate giant and pushes our vital concerns aside?"

George Crocker had been in jail during the battle of Stearns

County. He was extremely concerned about the use of base-ball bats and anhydrous ammonia, both on philosophical and tactical grounds. Crocker was anxious to get the protest back on the track of nonviolent resistance. New signs were hung on the walls of the Lowry town hall: "Martin Luther King was arrested 126 times. We also have a dream: No powerline."

From February 20 to 24, twenty-one protesters were arrested for using the passive resistance tactics of the civil rights and antiwar movements. They lay down in front of cement trucks; they blocked workers with sit-down demonstrations; and, making creative use of the materials available, they covered themselves with pig manure, blocked workers, and challenged the police to arrest them.

The pig manure action took place on Tony Bartos' farm near Constitution Hill. He and five neighbors rode out to a tower site on his land in a spreader filled with fresh manure. They jumped out and spread it on themselves. Bartos has recalled how it happened: "I had three neighbors and my brother, Darrell, and I thought we'd have to have one more, so I called Joe Zweig up by Lake Marion and asked him if he'd like to get arrested with me. He said, 'When,' and I says, 'Well, in about fifteen or twenty minutes,' and he said, 'Sure, I'll be there.' So we rode out there with a manure spreader full of manure, pigshit, you know, and we went out there and let them spread some on us. Then we walked down to the equipment and the cops wouldn't touch us. They offered to let us go three or four times and we wouldn't budge. Then they kind of threatened us with mace. We still didn't budge. So, finally, they didn't know what to do. I suppose they were calling all over. But anyway, finally they grabbed us and put us under arrest, took off our outside clothing, put us on a patrol car and hauled us to the road and then put us in an old bus that hauled us to jail."

Crocker's aim was to increase the civil disobedience, hoping to build up to mass arrests. At the same time, some of the protesters searched for leverage in other ways. Small groups went out regularly at night to test the vulnerability of the line. Survey stakes were pulled, concrete foundations damaged, and fences burned. The power companies hired more security guards, but it was clear they could not effectively patrol hundreds of miles of countryside.

George Crocker had become a leader at Lowry. He had moved out to western Minnesota to work full-time on the protest, renting a room at Mrs. Flynn's house, up the street from the Lowry town hall. By March, he was a highly respected voice in the daily meetings and a highly visible figure in the protest. Along with farmers Alice Tripp and Gloria Woida, Crocker frequently spoke for the protesters in the media. Many of the farmers felt that Crocker had more than proved himself by his philosophy of nonviolence.

About that time another leader was emerging. His name was Ben Grosz. An Alexandria native trained as an electrical engineer, Grosz had worked on the nation's only other +/−400 kv DC line, the Bonneville Power Administration's Pacific Intertie. He has described his work with Bonneville Power and the Bureau of Reclamation: "I was the man with the black hat for fifteen years. I was the one that had to go and keep peace with the ranchers and farmers as we went roughshod over them, condemning their property, mistreating the area and their land and what not, but when you have the right of condemnation and eminent domain law as it is today, the little man has nothing to say about anything. And that's really why I'm involved. I couldn't sit back and watch my neighbors in Pope and Stearns County just completely get bulldozed to the side by another big corporation, and that's exactly what it boils down to."

Soon after the protest began, Grosz moved back to the Alexandria area, opened a rock shop, and allied himself with the powerline protest. Because of his background, his expertise, and his oratorical skills, the tall, handsome Grosz soon found himself president of the Douglas County protest group, Save Our Countryside. In 1975, he had testified eloquently against the powerline at the state hearings. In 1978, he began coming regularly to the Lowry meetings. Because of differences over philosophy and tactics and differences in personality and lifestyle, however, Ben Grosz and George Crocker were on a collision course.

In March 1978, Grosz organized one of the most successful demonstrations of the entire protest. On Sunday, March 5, over eight thousand people took part in a "March for Justice." Following a tractor with an American flag, a coffin labeled "Justice," and an enormous papier-mâché figure, the "Corporate Giant," the people marched solemnly from Lowry to Glenwood to demonstrate their opposition to the powerline and their support of the protest.

Grosz has described how he got involved and what happened: "Sauk Centre businessmen said they were gonna have a rally that would put all our little rallies to shame. They knew how to do it and they were really gonna have a slam banger. It was gonna be at their cost, at their time. Our group had been looking for the last four years for a saviour to lead them to the promised land, someone to get this damn line stopped. A lot of people had been talking, including me, but none of us really stopped it. The Sauk Centre businessmen, they were gonna do it.

"Then a few days later Alice Tripp came to the Lowry town hall all disappointed. The businessmen had met to discuss how this was gonna be arranged and they got into a big fight. Half of them said they weren't supporting anything when it was headed by somebody that looked and talked and acted and had the background of George Crocker. The businessmen had dropped it, and man alive, everybody just gave up. The whole thing just fizzled. Well, I just said to myself, 'Holy smokes, we don't need a bunch of damn businessmen from Sauk Centre to do this,' so I just jumped up and I said, 'Okay, fine, we were all excited about a big rally and we will stay with it. At this time, we've been fighting this thing for three years. Either we have the support of the people in the area or we don't.' What better way to find out what support we have than to go on TV, give talk shows, get it in press releases that we're gonna have this big rally and here is the chance for all you people in this area to support a peaceful nonviolent rally to finally show the politicians of this state that we're not gonna sit back and have this thing pushed down our throat. And everybody said, 'Well, how can we do it?' I said we can do it the same way that they were gonna do it. I said, if you want me to head this thing, I will. I said I'm gonna need committees to handle these different things. Well, anyway hands started popping up, I just sat there and wrote down who had to do what and, by golly, next thing you know, I personally feel we had nine thousand people down here in Lowry. We had people come from as far as New York and California.

"We started out that day with church service. One was held

upstairs in Lowry and one was held at the Catholic church. After the church service, there was a mass meeting right outside the town hall where everybody was accumulating to get ready for the march and it was headed by a monstrous, big tractor, red, white, and blue, a brand new tractor. 'Justice' was there in a coffin and they were gonna bury him. The march from Lowry was several miles and it was a cold day; it was like twenty-one or twenty-two degrees below freezing. The idea was when we were about a half-mile short, then we were gonna stop the parade, let everybody catch up and then enter the campgrounds in force.

"Any time you get five, six, seven thousand people you've got a coordination problem. We started the meeting about a half-hour late. We had about a two-hour ceremony, and I limited everybody to about seven or ten minutes so we could get a lot of viewpoints. I didn't want one person up there monopolizing the podium for any length of time, and that was pretty well received by everybody. We had so many people there, we probably had to turn away twenty speakers from different areas. I was a very happy man. I don't think anything happened that day that we could be ashamed of—destroying property or messing up—we went out the next day and cleaned the whole site up; there were no beer cans, no mess, nothing. And we got our show of strength. Then we were gonna follow up on this show of strength with a march down to the Twin Cities the following Wednesday. We were all gonna meet down at the capitol and go in and raise holy hell with the governor and representatives down there."

Grosz didn't see arrests in the fields as doing much good; he thought fighting state troopers was a tactical mistake. If anyone was going to stop the line, it would be the governor and the legislature and that was where the farmers should exert direct pressure. Building on the momentum of the march for justice, Grosz planned intensive lobbying and a long-term sit-in: "If we could have had enough people, the more the better, if we could keep one hundred people there every day of the week—see, cause what we were shooting for was, they were just about ready to adjourn and we wanted them to extend the legislature one day to hear our grievances and to see if maybe we couldn't get a moratorium on construction of the line. I think they would have turned us down, but at least that would have been the only thing on the agenda that day and we would have gotten press coverage that would have been unbelievable. We might be sitting in the capitol for two months. We already had speakers picked out. These were regular farmers who wanted to tell how this land had been in their family for 110, 150 years, and what they were doing was wrong—give them a real gut feeling of what was actually in the minds of the people out here."

But first Grosz had a score to settle with George Crocker. That was one of the main reasons he had come to Lowry: "I got active again when George Crocker entered the scene, November of '77. Well, I guess if I had to label it priority one, two, or three, I would say that number one was to get the protest activities back to St. Paul where they could do some good, and then number two was to get George—who I didn't know, it was just his image, but I didn't like his image—get him out of there. And see, here again, George and I had never had a word between us, and I don't know the man, but I don't feel we needed him down there."

He explains why he wanted Crocker out: "George Crocker spent time because he wouldn't go over to Vietnam and fight, and for that I pat him on the back. But, I'll be very honest, George had the label of being a long-haired goddamn hippie up here. The farmers didn't like him. He had a following here now but the average farmer in this area can't stand the sight of George Crocker, and they felt he had another cause in coming up here. Shortly before the march, I spent most of my time defending George Crocker. . . . It got to be very old when I would spend half my time defending why good, hard-working, God-fearing people, farmers, would accept somebody like George Crocker. It's just prejudice in the worst degree, what a man looks like, but the papers did that. They would go out and protest and what would they get? It was George with his hair flowing, being dragged away, and you have to face it, there is prejudice in this world. . . . I personally had no objection to getting help from anywhere. Lord knows we needed it, but I didn't feel that any one of us should stand in the way of winning the battle, so to speak. I felt George Crocker was a liability as to what it was going to take to win this thing out here. It was going to be won by the governor stopping the line or the legislature, but as long as the average person in Minnesota felt that we were doing what George Crocker was dreaming up, that we were doing nothing but mimicking his ideas, we were never gonna get the support we needed in order to put the pressure on. . . . The farmers had a beautiful protest here. They had a hundred good reasons why this line shouldn't go through their farm. We didn't—they didn't—need somebody with a bad reputation to come up here and make it difficult to get the support that we needed to win this struggle."

Hoping to build on the euphoria following the march for justice, Grosz made his move on Friday, March 9. He went to the podium at Lowry and told the farmers there were problems they had to deal with; some hard choices had to be made. He introduced two motions. The first was that there should be "no more intentional arrests" in the field.

In support of his motion, Grosz asserted that the march for justice showed what a broad base of support the powerline protest really had. He argued they should adopt a strategy that would appeal to that broad base, and confrontation with state highway patrolmen did just the reverse. The resolution sparked a debate. Some still felt the protest in the fields and mass arrests were the best hope. But others, including some of the most influential leaders like Gloria Woida, were persuaded that the protest should move to a different phase. Grosz' first resolution passed 49–31. It was agreed that they would follow up on the march for justice with a mass demonstration at the capitol the next week.

Then Ben Grosz made his second motion—that George Crocker should leave the protest. He miscalculated on that one. Crocker had won many friends among the farmers in a few short months at Lowry. Grosz had gone too far. Alice Tripp remembers the reaction: "He was trying to tell that group, 'This is what you have to do,' and they reacted against it." Pandemonium broke loose in the hall. There was yelling and cursing and the meeting went out of control. Gloria Woida called for adjournment; a majority supported her and the meeting was over.

There never was a vote on George Crocker—but it was clear he had won the confidence of many of the farmers at Lowry. Tony Bartos has summed up how he felt about Crocker: "They can't believe anyone would come out of the Cities and help that's got a background like George's and wants to see the country run right. Well, anyone can say what they want about George. I think he is a nice guy and his ideas are what the people of this state and the people of the country need. And he isn't in it for politics. He could give a shit less—you know what I mean. They say, 'Get out of here, he is a communist.' Okay, turn around and ask them, 'Well, what is a communist?' They can't answer you. 'Is that somebody from Russia?' You know, they can't answer you." Many other farmers at Lowry agreed; they would support and stick by George Crocker, just as he had supported them.

The media picked up on the "no more intentional arrests" resolution. By that evening the whole state knew about it. Governor Perpich announced the next day that he was recalling the state troopers from western Minnesota. The confrontation was over.

The protest moved to a new phase, but it was not what Ben Grosz had in mind. He did lead a demonstration at the capitol, but only about a hundred protesters showed up and they were not well received. Grosz' capacity to lead had been severely shaken by his abortive attempt to oust Crocker. Even more important was an incident that occurred the night before the capitol demonstration.

Shortly before midnight on March 14, a security guard, Allen Krook, thirty, was shot at a powerline assembly yard near Villard, in Pope County. Fortunately, the bullet did not hit him directly, but he was injured by windshield fragments that showered him in his pickup truck. His assailants were not immediately apprehended, but it was assumed they were protesters.

This put a damper on the demonstration at the capitol. Ben Grosz recalls, "We were like a bunch of lepers down there." None of the legislators wanted to be associated with their cause that day. The shooting incident had a broader effect on the protest. It stopped the momentum of favorable publicity and public support that had been building since the march for justice.

Nevertheless, there was by then a solid core of protesters who were not about to give up, but who had few options left. The direction of the protest was about to change. There would be two last gasps "within the system," a serious attempt to implement the science court, and a surprisingly strong electoral challenge to Rudy Perpich—when Alice Tripp ran for governor. But the main thrust of the protest would move in an entirely new direction, really the only direction it could. What began that spring as evening "wiener roasts" around the towers would eventually move to something approaching all-out guerrilla warfare.

Math and Gloria Woida
Profile

Math and Gloria Woida live on a 320-acre farm, near where the powerline begins to head south outside the town of Sauk Centre. Among the most militant of the farmers, Math emerged as a leader during the confrontations with the state troopers and Gloria gained state-wide renown as one of the most effective spokespersons for the protest.

Why did you become so involved in this protest?

Gloria: Well, the reason why we got so terribly involved was that we really believe that people have to start standing up for their rights because people are being trampled on all over. Math's put a lot of long hard years into this farm and his dad, they bought it when it was absolutely nothing and built it up to a productive, meaningful business. And all of a sudden big companies come in and rip that away from you and put your family's life in jeopardy.

Math: I've been here since 1943. I come out here with my dad and mother when I was fifteen and I've been here ever since. My mother died the second year we was here and then I stayed with my dad. We first farmed the land with horses—didn't have the money to buy a tractor. Then we bought an old used tractor and worked our way up. And I put a lot of hard work in before we was married and after we got married, she helped do a lot of hard work, too.

When were you married?

Math: Twenty years ago.

How many children do you have?

Gloria: Four—they are nineteen, seventeen, fifteen, and eleven.

Gloria, did you grow up in this area?

I grew up about seventeen miles northeast of here. My folks were small farmers, so I've been on a farm all my life.

How many acres of land do you have?

Gloria: 320. It is a regular, general farm. We have sixty cows and do dairying and we also grow small grain and hay.

Do you like farming?

Gloria: I love farming. I could never be satisfied in the city. I like to be able to step out the back door and let out a holler. I mean in town I feel so confined and restricted. Out here I feel so free. I never had any ambition of being a highly educated person, which maybe I regret a little bit sometimes. I always wanted kids and I just naturally had in the back of my mind that some day I would be a farmer's wife.

Math: That's part of this protest. To keep the farm in the family, to protect the family farms because they really are almost becoming extinct, you know. You can't really get hold of land any more and it's almost impossible for a young farmer to start out unless he does have help from his parents, like handing the farm down to the next generation.

Have you ever been involved in politics before?

Gloria: I've never been involved in anything government before and I've really been kicked in my butt for that. I can see where it's because of the non-awareness in what's going on in the country, in the political system, that is why we're in the shape we're in. Would you believe I had never been to the state capitol before this protest? And Math was never involved either in government.

What lessons have you learned from this protest?

Gloria: Organize a group of people, the largest group of people you can as fast as you can, and start putting pressure on local government if you can. Put pressure very fast on all the levels of government that are affecting you.

How do you put pressure on—through the courts, the legis-lature, going to public hearings, disruption, or what?

Gloria: The fastest pressure is through protesting. We did a lot of running to the capitol, we did a lot of running to hearings, and it got us absolutely nowhere. So I think what the people should first of all do is put as much time and energy into making people aware of what's happening soon, fast, so you get a lot of public support on your side.

Math: And get a large number of people right out there on the road protesting. Keep the construction workers from ever starting surveying and stuff, even if there are arrests.

Gloria: And you have to watch the Energy Agency's certificate of need, where they're handing it out, and put the thumb on them right away before it gets too far. Keep an evil eye out on the agencies. I think we should have gone to those agency public hearings and just ripped the place apart, I mean not physically but just verbally.

Math: Physically would be a hell of a lot better.

With this early resistance, wouldn't they have just moved the line elsewhere, on other farmers?

Gloria: I don't know. I think a lot of other ones would have organized in a hurry. They would say, "Oh, gee whiz, those are such crazy fighting fools, they'll never put the line there, they'll probably consider us next." The power companies and the state might have had a heck of a time building the line. They might never have got the line built at all.

How about the courts?

Gloria: Total waste of time. We should have taken the money we spent on all of these court cases and got people out on the roads and fields. We got absolutely nowhere in the courts.

Why?

Because the court system is there just to back up government.

How about the legislature? Earlier, you said that you should

have been more involved in politics. Wouldn't this include vot-
ing and keeping up with the legislature?

Gloria: I really think that voting does carry some weight. I
think if the politicians feel threatened that the people won't back
them if they don't represent the people, this can help.

Math: I am not sure. It is the big companies and the bigger cities
that have the money and the majority of the vote, so I don't think
the ordinary guy, the small farmer, really counts too much. I still
say if we would have really put our time in organizing, and if we
could have built the group like on that one day we had one
thousand people out in the fields and they just lifted the surveyors
right out of the field with a chopper, they couldn't even protect
them with the troopers—if we could have had that many people,
they would not have gotten very far with this line.

Gloria: We'd have darn good support and then all of a sudden we
would decide to go run to St. Paul to the capitol. When that happened, that's when we'd lose a few people because they'd go down
there and feel so blasted frustrated, they'd come back with the
feeling, what the heck's the use of fighting.

Math: Like that one time [February 1978] we had the group
built up to a hell of a big number and they went down to the
capitol for some hearing or something—busloads and busloads of
people went down there. Well, then the people laid kinda low.
They figured the government was gonna do it; the governor was
gonna do something with the science court. After that it kinda
broke up and then it took a long while to get the people together.

But isn't it true that you were opposed to mass arrests back in
January [1978]? For example, when the state troopers first came
out to west-central Minnesota?

Gloria: I felt all along that a strategy had to be used where
we wouldn't turn the people off and they'd have to be seasoned
into the idea of confronting the highway patrolmen and getting
into peaceful arrests and having them accept that and yet coming
back again without the fear of being arrested again and try to keep
it nonviolent. The news media was just thirsting that first day the
troopers were out: "Oh boy, we'll have blood and they'll be head-
to-head the first day," and Rudy [Perpich] is thinking, "Oh, I've

got it now," and instead he turns on the TV that night and sees farmers handing out cookies, coffee, and flowers.

So you think that was a wise strategy?

Gloria: I wish, looking back, knowing what we know, I'd have sat down every day and been hauled off—that's the best way to protest and make a difference. More people should have got arrested, but they had to be seasoned into the idea.

Math: If we only could have had Ken Tilsen as a lawyer then. We didn't have as good lawyers then, and people were more afraid about what would happen to them.

Gloria: You know the first time getting arrested, the first time in your life—when I was arrested, I never had a traffic ticket or anything—it's almost a little bit like dying.

When you were named on that $500,000 civil damage suit, what was your reaction?

Math: I took it to a lawyer right away and told him I wanted to countersue for $600,000 right away.

Gloria: Our reaction was, by gosh, they're going to try to wipe us out, so let's go and give it all we got; they're after us, let's go after them. I thought we'd had it.

Math: I didn't think so. I went back out and protested. It didn't bother me, because I couldn't see where in the hell they could sue me for $500,000. We pushed to get the lawsuit into court and the utilities, they always had a reason for delay. So finally my lawyer said to Judge Hoffman, "The utilities only have a right to actual costs." The judge decided on $1296.18 if I ain't mistaken. Since then the companies said four or five times they'll drop their suit if I'll drop mine. But I says I'm not going to drop my suit against them unless my insurance company or the utilities pay the lawyer fees, because up to date it has cost me $3000. Why the hell should I pay it? And it never has gone to court yet.

Some people have said that the reason farmers in western Minnesota are involved in this protest is because the line is over their land and that is as far as the protest goes. It is a simple question of self-interest and no more than that?

Gloria: My reaction to that is yes, I used to sit home and be a

real quiet mama and take care of the kids and work on the farm—
just self-interest in your own personal little world, you know. But
all of a sudden the powerline comes around one day, and it opens
up your eyes to what people really have to do. I'm speaking for
myself; I can't speak for Math, but I will never shut up and be a
quiet person again and I will be involved in other struggles. Had I
known what the Indians were going through in Wounded Knee,
I'd have been there; and if I'd have known more about the Vietnam war protests, I think I would have been there.

Math: Well, there are some people who are gonna quit fighting,
but I don't think they should quit fighting, because there isn't
gonna be no farmland if you let these big corporations take over.
These power companies and stuff. The cities will be moved up in
the country and take up more farmland. The farmers are gonna
have to stand up and fight, not only in this area, all over.

Gloria: I wish people would realize that once government and
corporations control our land, we're controlled.

*When you talk about less government control, many busi-
nesses agree. Do you mean less government control of business?*

Gloria: No, less government control of just plain people's
rights. That is what we need.

Gloria, what has been the role of women in the protest?

Gloria: It really was kind of funny because it seems that some of
the really strong leaders at different times were women. Take
Alice—I admire that woman an awful lot. The first few times I
saw her in action organizing everybody, I said, "What a dynamic,
organized lady." And from watching Alice and starting to help
her, I started getting a little more forceful myself.

*Alice Tripp pointed out to us that women were not at all
active in CURE, that women were really excluded. But at Lowry
women seemed to be very active. How did this come about?*

Gloria: That just simply came about. When the guys were talking about something, like Virgil or Math, sometimes they would
have a really dumb idea and us gals sat there you know and we
started voicing our opinions and found that the guys were listening more to us. Sometimes I would come home and be almost

scared, because there were so many people coming up to you and asking you, "What do you think of this?" It was a lot of responsibility.

Sometimes Math and I would come home and we would disagree. Throughout our whole marriage, I had always given in to him. You know what I mean, if a decision was made, he made it, and I just learned not to question it. But I found myself during the protest getting the guts to stand up to him—doesn't it sound ridiculous. In a way, this struggle has been a great experience for me. My life has broadened so much and I never want to go back to that old role of being just a farm gal—you know what I mean?

The personal cost to your family has been very high. Your son faces charges and the both of you face misdemeanor and felony charges which could carry severe prison sentences. Then there is also the damage suit, which is not yet settled. Has it been worth it, being so involved in the protest?

Gloria: Yes. Would you have stayed out of it?
Math: I'm not done with it yet.

George Crocker

Profile

From the time of his antiwar and draft resistance activities in the late 1960s, George Crocker has been a well-known figure in Minnesota's radical politics. In spite of his youth, he has already been in the forefront of several movements for social change. Initially viewed with suspicion as an outsider, Crocker became accepted by the farmers as a stalwart in the powerline protest.

Why do you think there has been such a struggle out here in west-central Minnesota? Many people, for example, find it difficult to understand why so many farmers would be arrested for civil disobedience. Would they have resorted to these tactics if not for people like yourself out here organizing?

Just look at what they did before, like two years ago this fall up here on Scott Jenks' Constitution Hill. Just classic civil disobedience stuff. It was great, you know. Holding signs in front of the tripods, riding around on horses, spreading manure, doing all sorts of things of that sort that were very effective in getting the point across, but did not threaten the life or limb of anybody, and which stated very clearly how strongly farmers felt. They did this all before anybody from the Cities was involved with it. Of course, the more people you have looking at the alternatives, the more quickly you come to an understanding of what you can do. And I think people from the Cities did have an awful lot to do with that. But it's not that they brought something foreign in. They helped the people here refine what they already had, I would say.

I think the point that we're reaching all over this country is that it's getting real, real hard to be left alone. I think that's the critical factor. If people were left alone it would be different.

But they aren't going to be left alone. Too much price-squeezing going on, paying high to get the crop in the fields and selling low to the marketers, getting their land ripped off by powerlines, getting screwed in the courts every time they turn around, land prices out of control. I think if people were left alone they wouldn't fight. But this powerline struggle is an energy war. It is a battle against the increasing centralization of energy facilities and energy control. Western Minnesota is a sacrifice area just as is the Black Hills, Appalachia, and so on and so forth.

When did you hear about the powerline fight?

Well, I'd been keeping track of it pretty much since it started in 1974, just reading in the paper. Energy is a basic issue this whole society has to come to grips with, just like food. There are people struggling on food questions which was what I was involved at the time I was keeping an eye on this. People in the food co-ops— challenging the agribusinesses, challenging the chemical industry, its relationship to food production, all this. I've been here in Lowry since January 1 [1978].

Did you spend a lot of time in Lowry before you moved here permanently in January?

Yes. I probably came up from Minneapolis eight or nine times after November 22, when I was arrested. I came to the meetings in Lowry to talk to people. And some folks were saying then, "Well, we've just got to go and blow them away, that's the way we feel about it." I felt that if people had been killed, the farmers would not have been able to bring out their case to the people. The emphasis would have been, "They're shooting people, killing people," rather than "What are the companies and the state doing to the land, to the environment, the inhabitants?"

Why were the meetings held in Lowry?

When I first started getting really into it, most of the people I knew were from Stearns County and I wondered the same thing. But it was also true that one of the places that was available to us that wasn't a barn and wasn't a dance hall was the Lowry town hall, just about the only place that was close to the line. So there was a place, and of course there was a lot of support in the immediate area, and things picked up from there. Lowry was a real logical place to set our base and that's what happened.

How would you interpret the struggle out here?

This is an energy war. I think the policy that's being pursued is massive centralization of energy generation facilities and high voltage powerlines. That's what they're trying to do. And what this says is that you may get this one through; we'll take a bet on it, but even if you do, you're not going to get the next one through. People are learning too much. This line has blown a whistle on the corporate empire, just like the whistle is being blown at Rocky Flats, Seabrook, in upstate New York, at the proposed nuclear plant in Tyrone, Wisconsin. People are waking up.

Do you think what has happened here compares with Sea-brook?

I think what has happened here is in a whole lot of ways much more encouraging than anything that could possibly have come out of Seabrook, in that Seabrook was organized by people who came out of a movement background almost completely and then they developed their local support out of that—and they did a very good job of that and developing nonviolent tactics of confrontation. But it is the people they started with that made that happen, people with training of one sort or another and with a pretty developed political analysis. What we have here is the people who are being encroached on are the backbone of the resistance.

Let me ask you a few questions of a more personal nature about how you became involved in this. Are you from Minnesota?

Yes. I went to high school in Stillwater which is, of course, more in the eastern part of the state.

What do your parents do? Are they farmers?

Well, first, my family belongs to the Quaker church. In fact, my father was a conscientious objector during the Second World War. He went through an awful lot of heartache and trouble to be classified finally as co. At that time I think he was Congregationalist.

Anyway, he was finally classified as a CO and sent to public service camps in California and Montana. In both states, he was assigned to the Forest Service as a smoke jumper, jumping out of airplanes and putting out forest fires. When he was in camp, he got to know my mother who was a camp nurse. My mother was a Mennonite and Mennonites are one of the traditional peace churches. After the war, they got married and joined the Society of Friends.

What do your parents do now?

My dad is a carpenter and my mother is a nurse.

Do you have any brothers or sisters?

I have three brothers and a sister. My youngest brother and sister were adopted when I was in second or third grade. They're Native Americans. They've gone through a very hard time. My older brother is a Canadian citizen. He went to Canada with his wife rather than go to jail.

The papers talked a lot about your draft resistance and prison term. How did you become involved in the draft resistance movement?

Well, they tried to draft me.

Were you at the University of Minnesota when this happened?

No, I had only a few quarters at the University. I think I've quit college eight or nine times over the years. Anyway, I was traveling around—as far south as Tennessee and into that neck of the woods and coast to coast. Then the McCarthy campaign came along and I knew McCarthy was better than Johnson, so I went to Wisconsin to campaign for him. Then I was working in Indiana when my folks got word that the feds were looking for me for refusing induction. I didn't want to mess up any part of the campaign, so I just cut out of the McCarthy stuff and came back to Minnesota, and started working with the draft information center. And I worked with them from '68 through the spring of 1970, which was after I'd been tried and convicted and appeals had been denied.

How long was your sentence?

After conviction, the judge gave me four years; that was reduced to three years.

That was pretty steep for Minneapolis.

Yes.

Did you serve three years?

I did eighteen months and then parole. I was in prison in southeastern Michigan. It was a medium security prison.

What impact did prison have on you?

It reaffirmed my decision, that it was not a mistake. That kind of stuff—going to prison—either it kills you or makes you stronger.

And in your case, it made you stronger?

Well, yeah. There was every reason why it should. I knew why I was there, I was proud of the thing I had done that had brought me there, and I had a real strong community of support behind me. Not too many people going to prison have that kind of reassurance, but you have to have it.

During Christmas season, for example, I'd have so much mail coming in I didn't know what to do with it. Can you imagine getting over one hundred letters a day for three weeks in a row? You know, that was something, especially when other folks were locked up for the past five years and they ain't got a letter the time they was there. I didn't know how to deal with it.

How many other prisoners were there for draft resistance?

The population was about five hundred usually and I think there was twenty in for refusing induction.

Do you want to go back to south Minneapolis eventually? Is that where you consider home now?

My roots are in struggle.

What does that mean?

It means that you're prepared to stay here as long as necessary and then move on to other places. I'll be out in west-central

Minnesota till we beat 'em on the powerline and then I'm gonna take myself a break and I'm not sure what I'll do after that. I know there are some serious developments that are happening in the Black Hills with uranium, coal, minerals of various sorts, acid mining, solution mining, all that kind of stuff. They're gonna tear them apart if they have their way. I'll be involved with that. I know I will.

You were talking about going back to jail a few months ago. What do you mean by that?

For refusing to pay taxes. One of the reasons this powerline was able to go through is because people are willing to pay taxes; it's federal funds that are allowing corporations to do these things. If you look at the way the federal government spends its money, I venture to say that 90 percent of the tax dollars go to the ruling class—they are the primary beneficiary. I don't pay taxes, federal taxes; never have and never will.

So you are prepared to stay involved in these kinds of struggles for now and the foreseeable future?

With time off for good behavior.

What do you expect to accomplish?

We are gonna beat 'em, and if we don't, it won't matter cause there won't be anybody left to talk about it. I mean what we're talking about here is survival. Are we going to live? Will this planet be capable of sustaining life? I think people are strong enough to will their own survival. That means disobeying the courts. That means disobeying the vested interests of the ruling class. This is a movement and we're looking long-term. We're going to do it. Right now through this protest movement we're buying time with protest for alternative energy technologies away from nuclear and coal. We stop them and this creates a vacuum—we need to fill this vacuum with new ways of delivering and consuming energy. The issue is survival.

They want to take our dispute to a science court now

And judge if powerlines can hurt

Well, we've got more to discuss than a science court allows

We'll talk where we can be heard.

We've had enough of the government going along

With the money, whether it's right or wrong

They're getting too close now, you know they're making us strong

And they're gonna have to listen for once to our song.

—from "Minnesota Line" by Nancy Abrams

7 Who Sacrifices, Who Benefits, and Who Decides?

By the spring of 1978, when the troopers had been withdrawn, the farmers found there was only one thing left that they could legitimately talk about—health and safety. In the eyes of the public and, to an increasing degree, in the pronouncements of the protesters, health and safety became the central powerline issue. This was strange because health and safety concerns were not the dominant reasons behind the farmers' opposition to the powerline. Early interviews in this study make that clear. This, however, is typical of contemporary technology policy disputes. Witness the controversies over nuclear weapons testing, the ABM, the SST, and nuclear power. Somehow health and safety always becomes *the issue* in the public debates. As the powerline controversy illustrates, that is no accident. It arises from the underlying dynamics of public participation in technology decision-making—in particular, which avenues are really open to the public and which avenues are effectively closed.

One by one the farmers' concerns about the powerline had been bottled up by the state. From the beginning, the fundamental issue had been the question of need—the farmers were not convinced the proposed powerline across their land was really needed. The Energy Agency had held hearings and then issued a certificate of need. That made it official. The question of need had been considered by the state. It was no longer an issue.

The farmers were concerned about the effect of the line on an agricultural environment. The department of natural resources had prepared an environmental impact statement. That was all that was required. The question of environmental impact had been considered by the state. It was no longer an issue.

The farmers were upset by the way the power companies had taken their land. The EQC had held public hearings and selected a corridor and a route. In the process, all individual farmers could really say was, "Put it on someone else." But the question of siting had been considered by the state. It was no longer an issue.

The farmers were concerned about apparent irregularities and illegalities in the agency hearings—deciding on a corridor before considering whether the line was needed, transcripts which seemed to have been selectively edited, and so forth. The supreme court upheld the agencies on every count. That made it all legal. Administrative procedures had been reviewed by the state. They were no longer an issue.

The farmers were also concerned about possible health and safety hazards arising from the powerline. The health department had prepared a study. It was not definitive. The question of health and safety had been considered by the state. But it still remained an issue.

In this sense, the state processes channeled the farmers away from their other vital concerns and toward health and safety. The health and safety issue would not go away because, as is so often the case when new technologies are deployed, there was significant uncertainty about the risks. For example, as already noted, the literature on health effects of air ion inhalation had not yet reached a state of scientific consensus, and no careful measurements of air ion concentrations in the vicinity of a $+/-400$ kv DC powerline had been made.

The power companies' public relations departments did their best to deal with the health and safety "problem." They published newspaper advertisements with headlines such as NO LINK ESTABLISHED BETWEEN POWERLINES AND HEALTH, containing eye-catching pictures, tables of impressive-looking statistics, and reassuring rhetoric such as:

> In October of 1977 a Minnesota Department of Health Study concluded: ". . . there is no evidence whatsoever suggesting any effect on health or a sense of well-being from the inter-

mittent exposure experienced in the transmission line environment." . . . The Study was commissioned by the Minnesota Environmental Quality Board (MEQB) and represents one of the best summaries of current knowledge available on the subject of transmission lines and health. As in most areas of scientific study, the report suggests additional study of the subject is needed, but it does *not* recommend that any changes be made in existing transmission lines or lines under construction at this time. In short the report says that transmission lines are safe for humans and other living things.

In another era, the authority of the power companies or state agencies might have been believed. But Vietnam and Watergate and DDT and Thalidomide babies had produced a populace with a healthy skepticism of authorities and experts. The farmers were genuinely concerned about the hazards to themselves and their families of living near the powerline and viewed the power companies and the state agencies as interested parties whose word could not necessarily be trusted.

In March 1978, with much fanfare, the power companies had offered to take farmers and the press on a trip to the nation's other +/−400 kv DC powerline, just to see how safe it was. The Bonneville Power Administration's Pacific Intertie, running 846 miles through the Nevada desert from The Dalles, Oregon, to the outskirts of Los Angeles, had been in operation for eight years.

Fearing they would be used in a public relations stunt, the farmers at Lowry had voted not to participate. Only Virgil Fuchs and Werner Gruber had joined the expedition, along with thirty-six representatives of the news media and assorted public officials and co-op members. In two days on the West Coast, they were briefed by Bonneville Power officials and they had an opportunity to talk with some people living along the line.

Most of the people they spoke with reported no serious problems. But this was not really convincing evidence that the line was safe. There had been no systematic, scientific survey of short-term health effects among residents along the Bonneville line and, after only eight years of operation, long-term effects might not yet have become apparent.

Virgil Fuchs was a thorn in the side of the power company public relations people on that trip. In interviews with the press, he complained that the tour was rushed and tightly controlled; he compared their conversations with residents with asking smokers if what they were doing was safe; and he pointed out that the Bonneville line passed mostly over semi-arid, sparsely populated land, quite unlike western Minnesota with its many small family farms.

Owen Heiberg summed up his conclusions from the trip in the *Herman Review:* "The impression of most was there were no *immediate* dangers to health and safety, and this eased the minds of those who considered the line a necessity, but not those opposed, who returned from the trip confirmed in their opposition."

This was the situation that Governor Rudy Perpich found himself in during March 1978. For him, setting state troopers on Minnesota farmers had been a political ordeal of the most unpleasant sort. Perpich was very anxious to have the powerline dispute resolved. According to his aide, Ronnie Brooks, "There was nothing politically, socially, or economically we wanted to get rid of faster in the governor's administration than this particular dispute."

According to Brooks, Perpich had tried to deal with what he thought was really bothering the farmers: "We urged the co-ops to renegotiate, recontract and up their prices in fact, for compensation for people who had towers and lines across their land," and "modified the siting process . . . and included a greater weight being given to agricultural concerns and removed the primary concern for wildlife areas."

From the governor's perspective, the one remaining powerline issue was health and safety. In his January 1978 state-of-the-state message to the legislature, he had pressed for a science court to deal with that one sticky issue. Provided construction of the powerline could continue and only health and safety questions were considered, the co-ops had agreed to participate. The next move was up to the farmers. In March, Perpich announced he would meet personally with them to obtain their response. A meeting was set for March 19. Unaccompanied by aides, Perpich drove up alone to St. Cloud that Sunday morning to meet with a delegation of ten protesters at the highway patrol headquarters.

What he heard that morning surprised him greatly. The farmers' response was a novel counterproposal—a modified science court quite different from what the governor and the co-ops had in mind. As conceived by the presidential task force headed by physicist Arthur Kantrowitz in 1976, the science court proceeding would only consider the scientific questions relating to a controversial public policy issue. A panel of scientist-judges would preside as scientist-advocates interrogated expert witnesses and then the panel would issue its judgments on the scientific questions. What better way to deal with the vexing questions of health and safety that are so often the most troublesome issues in technology policy disputes?

From the farmers' perspective, there were still many power-line issues beyond health and safety. To be sure, public hearings had been held, but they were little more than an elaborate charade that had not seriously dealt with their concerns. Their counterproposal had been developed at a large meeting (which one of us [B.M.C.] moderated) in the Lowry town hall the night before the meeting with the governor. They agreed to relax their demand for a moratorium on construction and participate in a science court—provided certain conditions were met: the topics under consideration by the court would be broadened beyond health and safety to include need, alternative routes, eminent domain, impact of the line on an agricultural environment, and the missing transcripts mystery. The problem of a mismatch of resources was addressed; there would have to be a guarantee of adequate funding for the farmers to develop their case, bring in their own expert witnesses, and so on. The hearings would be directed at the public; there would have to be a guarantee that reports of the hearings would be widely disseminated.

Given the expansion of the topics to include nontechnical issues, having scientific experts act as judges no longer made sense. Nevertheless, the proceedings needed some closure. The farmers noted that since this was really a political issue, an accountable public official should make the judgment; they demanded that the governor not delegate his responsibility to "experts" and proposed instead that he himself be the judge. In the farmers' version of the science court, the governor's staff would monitor the adversary proceeding and then the governor would issue a public report stating and justifying his conclusions about the issues before the court.

When presented with this alternative proposal at the March 19 meeting with the farmers, Perpich expressed sympathy with the notion of including the question of need and with the necessity of guaranteeing adequate funding for the farmers to make their case. He was taken aback by the idea that he should be the judge, but agreed to consider it. Two weeks later, however, after conferring with his aides and the power companies, Governor Perpich issued a formal reply. He announced his refusal to broaden the scope of the science court beyond health and safety or to be its judge.

In his reply, Perpich clearly stated how his rejection of the farmers' proposal followed directly from *his* conception of the value of the science court procedure:

> The "Science Court" is a structure which is appropriate to deal with disputes over scientific fact. It is to be composed of persons technically competent in the relevant scientific issues, who can render some judgment on the state of scientific knowledge on disputed scientific issues. In this case, those issues relate to the health and safety effects of high voltage transmission lines. To broaden the scope of the issues before the Court or to have anyone other than impartial experts serve as "judges" would be to nullify the value of the Science Court procedure.

Replying to the governor a few days later, the farmers reiterated their insistence that the science court include the broad range of issues of concern to them. That was unacceptable to the governor and the power companies. There would be no science court in Minnesota. This attempt to bring about a "rational" resolution of the powerline dispute was dead.

Ronnie Brooks later put her finger on what went wrong, in a session on the science court concept at the American Bar Association's national meeting in August 1978: "Perhaps part of our failure was due to the attempt to impose a rational decision-making process on an already value-laden issue. If, in fact, the issue all along was one of intrusion on the land, then our attempts to address technological questions were at best irrelevant."

In retrospect, the contrasts between Governor Perpich's, the power companies', and the farmers' conception of the kind of science court that would be "useful" show clearly how inextricably such a forum is linked to politics.

While the health and safety question lent scientific overtones to this controversy, Governor Perpich was confronted basically with a political dispute. A politician's instinct in such circumstances is to hand over responsibility to someone else. And in that respect, the science court idea is a politician's dream—it focuses public attention on peripheral technical issues and delegates the decision to the "experts."

The power companies initially decided that they had nothing to gain from the science court—they had already won in the state agencies and in the courts. Later they apparently saw the science court as a means for diverting the energies of the farmers away from opposition in the fields; with no moratorium on construction, the line could be completed while the science court deliberations dragged on.

The farmers saw the science court in quite a different light. If restricted to health and safety questions, it would have been a fitting climax to the chain of statutory and procedural restrictions that had inexorably channeled them away from their basic concerns about the powerline—the sacrifice of their land without their consent for an allegedly greater social need whose validity they questioned—into a forum that would consider only quite peripheral technical issues.

The farmers' only hope of winning was to resist this channeling toward a narrow technical forum where experts decide and to keep the issues in the political arena. The critical elements of the science court idea, from the point of view of the farmers, were that they would be provided with *sufficient funds* to develop and make their case effectively and an *adversary forum* to face the utilities on an equal footing before the public.

The impasse in Minnesota reveals two defects in the original science court proposal. First, there was an unrealistic expectation of voluntary participation by all parties on the same terms in an *ad hoc* proceeding. Each party to this dispute made essentially political calculations as to what format, what timing, and what issues would best serve its own interests. It is not surprising that each arrived at different conclusions.

Second, the originators of the science court idea had an overly
narrow conception of what is wrong with our present processes
for dealing with technology disputes and this led them to an overly narrow conception of the kinds of new processes needed. In this
dispute, the farmers were channeled away from the issues they
really cared most about. Furthermore, although there was nomi-
nally full participation of all parties in public hearings about
the line, the power companies had adequate resources to develop
and present their case, while the farmers did not. As a consequence these hearings did not produce the kind of public discussion and debate of the powerline that would enable the people of
Minnesota to reach informed, independent judgments on the key
issues that underlay this dispute.

What might make a difference is something along the lines the
farmers proposed: an adversary forum to deal with the full range
of relevant issues, with adequate funding for all sides to present
their cases effectively and cross-examine the others' witnesses.

This kind of proceeding would represent a significant advance
over the formats currently employed in agency and legislative
hearings and the unsystematic and uneven information dissem-
ination processes that currently accompany state-wide referenda.

The failure of the science court marked the end of a year-and-a-
half exercise in symbolic politics. Rudy Perpich's populist style,
his visits to their homes, his willingness to listen, and his apparent sympathy for their position had initially raised the hopes
of the farmers. However, as his aide Ronnie Brooks remarked in
retrospect, "Expectations were perhaps unduly raised about getting some satisfaction from government." In the end, Perpich did
not bring the farmers satisfaction; he brought them state troopers.

The 1978 election gave the farmers another opportunity to express their displeasure. When Pope County DFL members met to
elect delegates to the state convention, they looked for a candidate to oppose Perpich. The obvious choice was Alice Tripp. The
confrontations in the fields around Lowry had produced many
leaders, but this fifty-nine-year-old former high school English
teacher, now a full-time farmer, had emerged as a chief strategist
and a principal spokesperson for the protest. Her courage, her
intelligence, and her open, direct style brought her respect in the
Lowry group. Her grandmotherly looks and her blunt, incisive

comments had made her a favorite among the correspondents covering the protest. By the time the troopers were withdrawn, television and newspaper coverage had made Alice Tripp a media personality, whose name and face were recognized in homes throughout Minnesota.

What began as a symbolic electoral protest grew into a full-fledged campaign for governor. The Tripp candidacy gained the support of a newly formed political alliance, the Farmer-Labor Association, and received a surprising 17 percent of the vote at the state convention.

Encouraged by this showing, Tripp decided to run against Perpich in the September DFL primary. One of us [B.M.C.] became her running mate as candidate for lieutenant governor. Dick Hanson of Glenwood was campaign manager and Patty Kakac of Alexandria was treasurer.

The two-month campaign was an ultra-low-budget operation, whose results exceeded the expectations of even the most optimistic observer. After the family pickup truck was outfitted with a solar-powered sound system (a small photovoltaic array coupled to a megaphone) and filled with gasohol, Tripp pledged that she would travel to every corner of the state. It turned out to be a grueling but joyful experience.

In part, it was just an old-fashioned political show. With Patty Kakac and troubadour Russell Packard singing and playing their guitars, accompanied by a huge papier-mâché puppet, labeled "Corporate Giant," the Tripp campaign appeared at dozens of county fairs and parades. The official campaign song, "Vote for Alice," composed by Packard, was heard all over Minnesota that summer:

Where you gonna be on election day?
Are we gonna let them have their way?
Pushing farmers off their fields to build powerlines
And nuclear plants their grand design.
Letting lawyers make the laws and judges say what's right
Or will you stand up and fight for your rights?

Chorus
Vote for Alice.
Vote for Alice,

Vote for Alice Tripp,
Vote for Alice,
Vote for Alice,
Vote for Alice Tripp!

 A vote for Alice is a vote to demand
Justice now thoughout the land
Guaranteed income, farmers' parity
Reduce the budget of the military
Freedom of choice for all women
Equal rights for all minorities
Tax the corporations, feed the poor
Deliver the wealth to every door
Public ownership of NSP
Power to the people, democracy!

(*Chorus*)

Mostly, however, the campaign was aimed at public education. A principal goal was to bring the substance of the powerline issue directly to the people of Minnesota—to explain the farmers' concerns, to describe how those concerns had been dealt with by the state agency processes, to demonstrate the inadequacies of those processes, and, hopefully, to bring about a reexamination of those concerns that had been swept under the rug by the state agencies.

For example, at one press conference Tripp charged that the certificate of need issued by the Energy Agency was "both illegal and inadequate." Illegal because the agency had failed to do its own independent forecast of energy demand, as required by law; inadequate because it had failed to consider the possibility of purchasing power from the MAPP pool as at least a short-range alternative to the CU project.

At the same press conference, she reported that Robert Banks, who headed the health department study, felt that there were two possibly serious hazards, air ion inhalation and electric shocks due to transients, that could and should be investigated before one could say with confidence whether or not the CU line was safe. She suggested to reporters that they would get an important story if they interviewed Banks.

The 1978 DFL platform, adopted by the state convention, had called for a moratorium on construction of the powerline.

Throughout the campaign, Tripp called on Perpich to declare publicly whether or not he supported his party's platform on the powerline.

The campaign was by no means limited to the powerline issue, however. Tripp tried to explain to the people of Minnesota how what happened to the farmers was symptomatic of a much more general problem that affected all citizens. Wherever she went, Tripp posed the question: "Who sacrifices, who benefits, and who decides in America today?" She began her answer by describing the farmers' experience with the powerline: "What the farmers found about the powerlessness of ordinary citizens raises disturbing political questions: who really controls our country and who will shape our future? The farmers found their land being taken away by an imperious and impersonal process that had no respect for their concerns. What truly surprised them was the role of their government. The utilities had planned the powerline; from the beginning the governor and his agencies worked to help them build it."

She related this experience to the general question of who decides in America today: "What happened to the farmers of western Minnesota is happening to all of us. In energy, in agriculture, in health care, in almost everything that matters to people, the same pattern of power is evident. Partnerships of large corporations and entrenched bureaucracies dominate the vital decisions that affect our lives and shape our future. Corporations call the tune; the government—our government—helps them carry out their plans. This is not democratic. The corporate planners and government bureaucrats who decide policies are not accountable to the American people."

The key issue, Tripp concluded, was the accountability of those who make policy: "Decisions that affect our lives and our future must be made by the people themselves or by individuals accountable to the people, and the people must know when these decisions are being made and have access to sufficient information to allow them to form their own independent judgments."

Press conferences arranged in major cities around the state provided a platform for nonestablishment people to present nonestablishment proposals for dealing with issues related to energy policy (natural gas pricing, a comprehensive decentralized solar and conservation program), welfare reform, women's rights, and the responsibility of corporations to their communities. All these issues were related to the campaign's central theme of the role of government vis-à-vis corporate interests and ordinary citizens.

Between a political campaign and its ideas and the public it seeks to reach with those ideas stand the media. Lacking funds to buy television and radio time or newspaper space, the Tripp campaign was almost totally dependent on media news coverage. Aside from state-wide public radio, which broadcast several lengthy reports on the substance of the campaign and the press conferences, and local newspapers and radio stations, which were eager to interview Alice Tripp when the campaign came to their town, the media was a huge disappointment.

The major sources of news for most Minnesotans are television and the major metropolitan daily newspapers, foremost among which is the prestigious *Minneapolis Tribune.*

It is one thing to *know* that TV news coverage is superficial because only a minute or so is devoted to any one subject on the evening news, but it takes standing in front of the cameras with the naive intention of saying something substantial really to appreciate it. At her first press conference, Alice Tripp decided to air her position on women's rights, the ERA, and employment opportunities for women in government, and there was a natural time to do it—August 26, the fifty-eighth anniversary of women's suffrage. It seemed appropriate to hold the press conference in the rotunda of the capitol in St. Paul and have her begin by pointing out that while it was fifty-eight years since women got the vote, there had never been a woman governor of Minnesota; in fact, neither major party had ever endorsed a woman candidate for that office. To entice the media to attend, it was announced beforehand that Alice Tripp would speak on "Minnesota Governors and Sex." She planned to follow her opening remarks with a detailed discussion of the dismal record of employment for women by the governor on his personal staff and in his state agencies and conclude with her position of the ERA and women's rights.

From the beginning it was clear that this conception was all wrong. The television people just wanted a shot of Alice in the capitol rotunda, surrounded by pictures of former (male) governors, saying there never had been a woman in the statehouse. A discussion of women's issues or a presentation of facts and figures on employment in state government was clearly beyond the depth of this forum. With a minute-and-a-half of good footage in the can, the lights went off and the television reporters were ready to leave.

At least they came. Most of the major newspapers never bothered to attend the press conferences. For a campaign aimed at public education, this was a fatal blow. Newspapers were the key to communicating the campaign's ideas to the public and to establishing the campaign's credibility. Especially important was Minnesota's premier paper, the *Minneapolis Tribune.* However, the *Tribune* failed to cover any of the policy proposals and any of the issue-oriented press conferences. It published only one substantial news article abut the campaign in two-and-a-half months and that was almost totally frivolous—it lumped Alice Tripp together with eight fringe candidates for the Senate (whose total vote in the primary later turned out to be about equal to Tripp's alone), saying "their campaigns range from the quixotic to the virtually nonexistent." This lack of coverage was puzzling; Alice Tripp was a woman who was well known in the state and who was making a serious attempt to relate her experience in the powerline controversy to issues in the lives of other Minnesotans. There was clearly a story there, and for whatever reason the *Tribune* failed to cover it.

Four days before the election, the *Tribune* finally did publish a serious page-one story about Tripp and it was an old-fashioned smear. Headlined POWERLINE PROTESTERS HAVE INVESTMENTS IN COAL, the unbylined article reported that the Tripps and some other farmers involved in the powerline protest had invested in an Iowa coal mine and implied that they stood to profit if the CU powerline were stopped.

What really had happened was at worst a case of questionable judgment; in 1975, lawyer Norton Hatlie had suggested to farmers in CURE that they not pay him directly for his legal fees, but instead invest in some of his business ventures, including Hawkeye Mineral Resources. Several of the farmers, including the Tripps, had given him money for this purpose, but the notion that their opposition to the powerline had anything to do with personal financial gain from an Iowa coal mine was simply untrue and every *Tribune* reporter who had covered the protest knew it. The article had evidently been in the works for some time; a reporter had called Alice Tripp several weeks before to ask for her reaction. That it was published as a page-one story on the eve of the election was surely no coincidence. A press conference to reply to the charges was hastily arranged, but few reporters chose to come out on a Saturday morning. A short report in the *Tribune* was buried in the back of the Sunday paper and other outlets which had picked up on the *Tribune*'s sensational charge had not a word about Tripp's explanation.

Three days later, on Tuesday, September 12, the voters went to the polls. The result was surprising to say the least. Running against the incumbent governor, Rudy Perpich, Alice Tripp shocked everyone, including herself, with over ninety-seven thousand votes, 20 percent of the total. With a total campaign budget of under five thousand dollars, it cost about five cents a vote, close to a record in that department.

In the limited sense of electoral tally, the Tripp campaign was a success. It provided a real shot in the arm to the powerline protesters, demonstrating that there was still substantial support for their cause around the state.

In the broader sense of raising issues and educating the electorate, the campaign was a failure. For example, it failed to arouse any significant support for a serious reexamination of powerline issues. Tripp called on Governor Perpich to indicate whether or not he supported the DFL platform plank calling for a moratorium on construction. Perpich ignored her. No one in the media pressed him to reply. Tripp announced that the man who led the health department study now felt there were serious health and safety questions that deserved immediate attention. No one in the media went to see him. Tripp charged that the certificate of need process had been illegal and inadequate. However, no reporter took the trouble to investigate whether or not her allegations had merit.

In short, it was like shouting into a void. In response to Tripp's complaint, she was granted an interview with the *Minneapolis Tribune*'s managing editor, Frank Wright. That turned out to be an exercise in futility. He listened politely as she described the almost total blackout of reports in his newspaper about the serious issues she had raised in press conferences around the state. Admitting it might have been a mistake not to cover her campaign, he explained that the decision had been simply a matter of "news judgment." So much for Alice Tripp and her educational campaign, as far as the *Minneapolis Tribune* was concerned.

In retrospect, by working within the system the farmers were doomed to failure. It was not because of a lack of opportunity to speak. The siting hearings and the need hearings had brought plenty of opportunities for "citizen input." It was not because all the public officials were unsympathetic to their concerns. They were unlikely ever to do better than a populist governor like Rudy Perpich, a liberal aide like Ronnie Brooks, a progressive energy director like John Millhone, and a fair-minded hearing officer like William Mullin.

They were doomed to failure because in calling for *no powerline* they were calling for a fundamental change in the traditional way of doing energy business. Such fundamental change in the system does not happen merely as a consequence of rational dialogue or fair-minded officials. The reforms that led to the creation of the EQC and the Energy Agency were system-supporting reforms—they did not much change the pattern of power or the basic assumptions in energy planning. When all the hearing records are sorted out, it is clear that the farmers' contribution to a rational dialogue was effectively limited to pleading that the line be put on someone else.

"Who sacrifices, who benefits, and who decides?" From the farmers' perspective, in the case of the powerline, the answer was clear. They sacrificed so that institutions like UPA and CPA could thrive and the insatiable energy appetite of an energy-fat America could be fed. The utilities planned the powerline; the government acted like their partners in the project. The farmers felt like outsiders looking in while government planners assisted corporate planners in carrying out their program. When the power companies encountered a hitch, as in Pope County, the state government was there to overcome the opposition.

An obvious question is why government does so often act like a partner to corporate interests. The answer is certainly not simple. In part it has to do with shared assumptions among planners in government and in business. In addressing this question,

Arthur Sidner, who replaced Peter Vanderpoel as head of the Environmental Quality Council, emphasized the existence of a
"natural fraternity" of professional planners. In Sidner's words:
"Why do government and business come to get together? One
reason is that most of the decisions are made by 'professionals.'

"These professionals have a common culture. I was trained at
Washington University to go to work for McDonnell Aircraft, not
to be self-sufficient. So my simple explanation is, when people
get ready to make a decision, they tend to relate to other people
that they feel comfortable with. And I think it's simply the common cultural experience. That's why they talk to each other and
that's why they feel comfortable with making that decision that
their friends will accept. . . . It's a natural fraternity; they have a
natural fraternity.

". . . What I'm saying is that those people are tremendously busy
people; they don't have time to be involved with everybody. They
make decisions the easiest way they know how. And that's because somebody they trust told them that's a good thing to do."

Thus, contributing to the development of a natural fraternity is
common training which emphasizes common values and long-
term professional relationships among people who have worked
together before and expect to work together again. Also contributing is the well-known "revolving door" phenomenon, where the
same planners move back and forth between government and
business. It is not surprising to find Larry Hartman working for
Commonwealth Associates one day and the Environmental
Quality Council the next. A man who helped plan an NSP corridor
through western Minnesota for Commonwealth Associates has
just the kind of expertise and contacts needed by the EQC to site
the UPA-CPA powerline. Nor is it surprising to find Peter Vanderpoel moving from head of the EQC briefly back to the *Minneapolis
Tribune* and then on to his present post as chief of public relations
for the Northern States Power Company. His insider's understanding of the state siting process and his contacts in government and the press surely can mean a great deal to an electric
utility like NSP.

But they also can mean a great deal to ordinary citizens like the
farmers in dealing with their government and with the press. The
farmers were not a part of that natural fraternity of planners and

they knew it. They knew they were up against people who did not share their rural values; they suspected that the public hearings they attended were mostly window-dressing, while the important dialogue took place in private conversations among planners. Such private dialogue did not necessarily have to be in the nature of "collusion" in an invidious sense—it was more likely just some paid professionals with similar assumptions and values talking together about how to get the job done.

A sad lesson of the Minnesota farmers' experience is that changes in basic assumptions and values will not happen solely as a consequence of playing by the rules within the system. The rules have to be changed, and that necessarily involves going outside the system. The massive civil disobedience in the fields during the winter of 1978 was a first step. More effective ways of making the system respond would soon be discovered.

Dick Hanson
Profile

Dick Hanson lives with his parents in Glenwood, the Pope County seat. He works on the family's dairy farm and is also a student at the University of Minnesota, Morris. He is a social activist and an astute political strategist. Active in the resurgent Farmer-Labor Association, he managed Alice Tripp's campaign for governor, served as a delegate to the 1980 Democratic National Convention, and has been elected as one of four Minnesota representatives on the Democratic National Committee.

How did your family get into farming?

My grandfather bought the farm over fifty years ago and my dad farmed with him after he graduated from high school. Then the year after he got married, there was a tornado that came and blew down our barn and windmill and everything. The farm was pretty much devastated and then at that point my grandfather collected insurance from it and my father bought the farm from him in that condition and took it over in 1946. It was 183 acres. Then my dad rented several hundred acres. He kind of went into machinery—he was very mechanical—he was one of the first ones to get a cornpicker going. And then there were a few years during World War II when the grain price was good and so he was able to buy more land. By 1963 we had 473 acres. We've been in dairying almost all the time. It started with a couple of dozen cows and then building onto that. I started milking cows when I was a second or third grader—that was about twenty-three years ago—and I've been milking cows ever since.

Does the powerline cross your farm?

The original route the utilities wanted went right directly across it. They switched it several times. Now the line is twenty miles or better away from us.

Actually, when we knew it was crossing our land, I guess my dad kind of had the feeling, and I did to a certain degree, "Well, let's not be so hasty in opposing the line, if it is for the good of the many."

We didn't really know much about it. They first came and started surveying and they wouldn't tell us who they were. My dad went down and asked them what they were doing; were they doing some more road construction? And the guy absolutely would not say a word. We found out what was going on through a cousin and I guess the first thing we did was start attending some of the public information hearings. The farmers were treated like shit. The whole process was rotten; they never answered a question straight. Then my dad and I went around with a petition against the line and there was not a single person who turned us down. Then the line got moved to the Lowry area and it was interesting to see certain people drop out at that point, since the line was no longer on their land. But I would say a good portion of the people who were affected at any one time stayed, no matter where it was put.

Did you go into farming right after high school? When did you graduate from high school?

I graduated in the spring of '68. I was accepted at the vocational institute in Alexandria, fifteen miles from us. I attended their agricultural salesman and marketing program. It was an eighteen-month program. In the middle, I did job training with Sears and Roebuck, getting experience in retail sales. I graduated in the spring of 1970. About halfway through, I guess I wished I had gone to college.

Why didn't you go to college?

I guess I didn't think I could even get in college; my grades were low. I assume they were low. It depends. My last year's were good.

Why the low grades?

I just set real low goals. And then I became very active in the

National Farmers Organization (NFO) and it was a whole family thing—attending meetings and rallies. Dad spent a lot of time organizing and so it left a greater burden on us to carry the farm load. I looked at authority in a real rebellious manner and took that up in high school. Also, quite a number of teachers in school were very negative towards the NFO and the farmers' organizing. I remember after we lost this one holding action, one teacher right in class, it was in geography class, said, "Well, now that you NFO people have lost, when are you NFO people and niggers going to go back to Africa where you belong?" The community was real split over the NFO, and for some there wasn't a four-letter word that was as bad as NFO, so it was kind of like being a "nigger" or something like that, at least that is the way I always felt.

How did you feel about the Vietnam war?

My opposition to the war grew during high school and I worked for Eugene McCarthy. There was a pretty strong McCarthy movement here. We had six delegates to the [DFL] state convention that voted for him. It was a combination. In 1967 we in the NFO had a milk-holding action where we dumped our milk and we were very confident we were going to win our goals. Then Lyndon Johnson got a federal injunction against us to stop—we had to sell our milk, and that destroyed us. There was a very bitter feeling toward Johnson, and so we had kind of an interesting coalition of old crippled-up farmers aligned with the young antiwar group. I would say most of the NFO people were antiwar, too.

When I first registered with my draft board in 1968, I checked out getting CO status. I knew at that point, I could not go to war. Well, at that point it was the Vietnam war; I hadn't really thought through about *any* war. The person in charge of the draft office said, "Well, unless you are a Jehovah's Witness or Holy Roller or something, forget it." I didn't give it much thought until after I graduated from vocational school. After graduation I worked on the farm. My two brothers were both in the service and there was a real need for me and so I got a one-year agricultural deferment. During this time the supreme court ruled you could oppose the war on moral grounds, not just religious grounds, and that gave me hope. So then I applied and the board turned me down. But legislation went into effect at that time which said that anybody who had served for twenty-five years or longer had to resign or be replaced and it seemed like it got rid of most of the draft board in our county. I appealed and the new board approved by a 3–1 vote. I was real sincere. I told them I would go to jail before going to war in Vietnam.

Where did you do your alternative service?

In California. I was in the California Ecology Corps. Ronald Reagan had set this up with big publicity; he was getting the co's to work on environmental things. It sounded real good on the surface. The local draft board told me about it, and I went out there in January of '72. It was a very interesting and exciting group of people, but the conditions were very bad. The reason Reagan had set this thing up was not the high and mighty thing that he was concerned about the environment. The backbone of the fire-fighting unit in California was the prisoners, the convicts. Well, over the years the judges had become more liberal and the number dropped way down—they paid them fifteen dollars a month, real cheap. So he thought this would be good for these no-good-for-nothing, trouble-making co's, and since the prisons were already there, all's he had to do was take the bars out.

How long were you there?

I was out there for about seven months.

What did you do when you came back to Minnesota?

I immediately started organizing for McGovern and became Pope County coordinator for McGovern. We set up an office for two months, one of the best offices in the state. We recruited about one hundred people and we wrote personal letters to farmers and door-knocked. We put in a real effort and carried Pope County and the other counties in west-central Minnesota—they went for McGovern over Nixon.

Is Pope County in general a Republican or Democratic county?

It varies. It is a rather independent group I would say. It can vote Republican, but then it elected some of the Farmer-Laborites. The first Farmer-Laborite to run for Governor in Minnesota in 1922 was a dentist from Glenwood and in 1924 he was elected U.S.

senator and served for thirty years. This area was strong for Floyd
Olson and Elmer Bensen, who was governor from 1936 to '38,
comes from the county just south of here.

What did you do after the campaign was over?

I made the decision to give school a try. I had been accepted at
the University of Minnesota at Morris. I was very disappointed in
the total apathy of the students; everyone just seemed interested
in getting in whatever was the best-paying job. Maybe it was just
the times. Actually, I knew more of the professors than students.
It just happened at this time the NFO had embarked on an organizing program and the state president asked me to work with
them. So I dropped out of school in '73 and organized with the
NFO. I became very frustrated with the national board and resigned after a year. The state board supported me, though, and I
stayed on and worked as a lobbyist for one more year in Minnesota.

Is it true that you gave a thousand dollars to the NFO?

Well, in 1973 the three largest dairies sued NFO for twenty-five
million dollars. It was a deliberate effort to break us. So we had to
raise funds for legal work or the organization would have gone
under completely. I was also very frustrated at this time with the
national board and had resigned. But I did have this money I had
been saving up, so I gave everything I had saved up to that point to
help out—it was a thousand dollars.

What did you do after 1974?

In 1974 I worked in Bob Bergland's campaign [Bergland retained
the congressional seat and later was named secretary of agriculture in the Carter administration]. I had supported him the past
two elections. We also ran several senate campaigns to the state
legislature in '74 and '76. One of them was a big upset. Roger
Strand, a recent graduate of the University of Minnesota, Morris,
beat a real reactionary Republican.

I also ended up working for Carter and Mondale, too. I didn't
like them, but when some of the liberal groups like the ADA came
out and supported them, I felt like I had to help.

Do you mean during the general election?

Yes. My choice for the nomination was Fred Harris.

Let me ask you about the arrest at the Rutledge farm in January 1978. What did you hope to accomplish?

It was in a sense making a political statement. Maybe not as strong a one as when you go and saw a tower down, which I believe is making a political statement, too. But I was hopeful, and still am, that the example would cause other people to get arrested, to put massive pressure on the government and the companies. Martin Luther King was a very important person to me by the example that he set and what he was able to accomplish through nonviolent civil disobedience.

How were you familiar with Martin Luther King?

They had done a special on King; it was NBC that did a special which showed all the number of times he was arrested. And I remember talking in front of the group at Lowry town hall and asking, "How many people saw it?" Probably three-fourths of the people had seen it. And just about everybody identified with it; they were moved by it.

What do you think is the future for farmers? What kind of policies could help family farmers?

There are a couple of things that could really do it. I have a feeling myself that cancer, which is one of the biggest killers in this country, is caused by chemicals and that these chemicals are being used with ever-increasing frequency on our farms and that this is directly linked with this cancer. Organic farming—rotating crops, balancing things off through nature, rather than using more chemical fertilizer—this, in a sense, requires the small farm, that you have the livestock, the hay, the crops. And there are more and more farms that are becoming organic farms. I think two hundred to three hundred acres or smaller is the ideal size. And if people become conscious of wanting to have the food they eat raised more this way, then that might save the family farm and bring more people back into farming.

I think that grain alcohol has great potential. Farmers right here in Minnesota could be energy-independent exporters of energy through grain alcohol, methane, wind power, and large sources of biomass. I see this as the salvation of the family farm and I am very excited about that. I am just worried that the oil companies will get ahold of these renewable sources of fuel. I am really worried that if credit is cut off—our banks are saying they're out of money—if the farm credit is cut off, we could lose three-fourths of our farmers. They would just go under and we could have just a complete corporate takeover.

What are your personal plans for now?

My decision is finish college—it is very important to me. I should be finished next year.

Then what?

I expect to be socially active. Maybe in government, if it would do any good. I'll work in some way and be an activist. Maybe underground or maybe above ground. I'm going to follow what I feel.

Patty Kakac

Profile

Patty Kakac is a young sensitive farm woman from rural Douglas County, with a deep commitment to the land. She speaks softly but with great feeling about the struggle of family farmers to survive in America today.

How did you first get involved in the protest?

Well, I was interested in what was going on. I started reading about it in the paper, reading letters to the editor about how terrible the line was that was coming through.

What paper?

Well, our local paper, which is from Alexandria. It drew my interest. Things like a big powerline were bothering me for a long time, you know, things coming into our rural environment, destroying natural beauty and farms. So then I was just keeping up with it in the newspapers and I went to a couple of meetings.

Were these meetings before or after January 1978?

Before then. I don't remember exactly when it was, probably in '77.

So you were involved back in 1977?

Well, not much. I never wrote any letters to the paper. I was just reading different people that would write in about it and hoping that it wouldn't come near our place. Just like everybody else. And then, I guess what really made me think I should get involved was when I heard about the arrests going on. Farmers and their wives getting arrested. I thought, something was really crazy, that they'd be arresting these people, seeing the troopers grabbing these people.

This is after the state patrol came in? You are not talking about the arrests prior to that.

Yes, this was after the troopers were called in.

Was there any particular arrest that upset you?

I think the Rutledges'.

Why was that?

I guess, hearing that a man and his wife were both of them arrested right on their own land, that bothered me.

Did you know any of the people involved in the protest?

No, I didn't. I live closer to Alexandria, close to freeway 94. That is pretty far from the line or where Lowry is. But one day my brother and I, we were talking, you know, at home all the time about what was going on. Finally, I said, "Well, we shouldn't be sitting here just talking about it. We should see if we could do something to help." And then one day we drove to a meeting in Lowry. This was really a short time after the Rutledges were arrested. Actually, I remember now, it was two days before that Friday when they got five hundred people in that hall, they were afraid the floor would go through there were so many in the hall. Anyway, we talked to Gloria [Woida] and told her we would do whatever we could to help. And Gloria said, "Oh, we would love it, whatever you can do to help."

Did you get involved in any of the arrests that were to take place?

No, I just went to meetings after that and kept pretty quiet.

What was your feeling about getting arrested?

I thought—I respected those that were arrested, but I was scared of it. They were just getting to the point they were asking people to volunteer for arrests, you know, a group of people to go out and get arrested. That was just shortly before the March 5th rally. And I was just at the point I was going to volunteer to get arrested.

Then Ben Grosz said that we shouldn't have any more of these arrests until after the rally.

That day when Ben Grosz introduced a motion that there should be no more "intentional arrests" and you should focus on the legislature, what was your opinion about that?

I had a lot of respect for Ben, so I thought maybe he had a point. But I disagreed with him, because it was the arrests that had brought me out and it was the arrests that were given publicity.

But you said that you were kind of scared of getting arrested. Why was that?

Well, I was afraid of, you know, what relatives and friends would think; that was part of it. Part of it was, I suppose, being raised, you know, that you don't go against the law and people getting arrested were lawbreakers. I had never been involved in protesting before.

What do you think was the best chance for the farmers—the mass arrests, the courts, the legislature, or what?

Well, looking back on it, there's lots of things that might have been done, but I don't know it would have changed that much. 'Cause now that I know about it, it's a pretty big thing to be fighting, because it isn't just involving these two co-ops, UPA and CPA. The whole country is interested in it and they would be affected if this thing was stopped.

What do you mean?

The powerline could affect the whole energy problem all over because if we stopped this line—of course, it depends on whether you consider stopping it or moving it someplace else, but that coal out in North Dakota is one of the biggest things for energy now. What the nation, what the oil companies are looking for is that coal out there and they wouldn't want to see this line stopped.

Earlier you said that one of the reasons you became involved in this protest was that even before the powerline you've been very concerned about the rural environment. I never asked you what you meant by that.

Well, it's hard to explain my feelings, I guess, about the earth. I suppose, it's been kind of instilled in me, maybe through my father, how important nature is to us. And I suppose part of it is being the youngest in a family of ten. I was kind of a loner, because my parents were pretty poor when I was being raised, and very busy. I was kind of left to myself, so I would go out and enjoy nature a lot. And as I was growing up I could take a walk through the woods and there were the wildflowers and wild berries and we lived near a lake. And then I could see the houses were going up all over, trees being taken down, all the stuff that I loved so much, the wildness of it being destroyed. And farmers around, you know, small country farmers, I just have such a deep feeling for—they were having to move away, selling to people from town or selling to big companies who were buying large acreages of farms, and it wasn't the small farms any more. It isn't like our rural environment and that really bothers me. So I started getting interested in organic farming and organic food and I started reading about and finding out about nuclear power. I'm really against that, too.

Do you still live on the small farm you grew up on?

We lived in a small little shack in the middle of the woods. Actually, this probably has something to do with my protesting, too. It was mostly woods and we had probably thirty to forty acres of field and we had a couple of cows and some pigs and for my whole family the big dream was to build it up and have a bigger farm, you know, but we were really poor. One day we came home from town, here surveyors had driven through our strawberry patch and my mother and dad couldn't figure out what it was, what they were doing out there, driving right through our garden. Then they found out a highway was going right through the middle of our place and they were going to condemn it. My folks fought the condemnation. They got a little bit more money. People thought they were funny people to be fighting a highway that was needed so badly. They talked about how bad it was that they cut all those trees down for the highway, and used up good farmland. This is what my parents said. I was about six then. We moved to this place south of Alexandria when I was eight. Well, we started on eighty acres and started building up, but then the barn burned when I was sixteen, so then we moved to another farm nearby, which is the farm I live on now. It's a dairy farm. They have about six or seven hundred acres that my brothers own, and they rent a couple of hundred. They milk sixty cows.

Do you farm? Somebody said you sing and somebody else said you were an artist. You are a lot of things. What do you do?

I do lots of things and not much of anything.

What are the lots of things you do?

Well, see, after high school I went into nurses' training and nursed for about a year-and-a-half in Montana. Then my mother was getting sickly because she had diabetes and she needed help and she was going to have surgery, so I came home. I started helping my mother and living on a little farm place a mile-and-a-half from our farm. It was a vacant house and my sister and I went over and fixed it up to live in. And then I had a cow and pigs and chickens there that I took care of. And then I would go home and help my mother. And I did that till my mother died. At that time my brothers were adding on to the house, so they asked me to come home and help with that and help on the farm. So I renovated the house and helped take care of the horses, the chickens, the garden and yard work, and cooking and washing clothes and house-keeping.

When did you start singing? I know you sang in the campaign for Mike and Alice.

I never sang before that—there's been a long time eating pow-dermilk biscuits to get the courage to sing in front of people. When I was in nurses' training I started learning the guitar, and then it was so painful and always I quit. And then I just played a bit more and more on my own, just for my own pleasure. Then I would sing before people when my mother and relatives would ask me to sing. But I never sang too much for groups of people.

Are you going to stay involved with the protest?

Right now there is much to do. We're at a point where people who are fighting this have to come to realize if they want to continue the fight against the powerline, it has to develop into something broader. People have come a long way in their thinking and working but they're still not at a point—they're fighting to get this powerline down, away from here, but it has to be more than that—it is part of a whole energy struggle. People have to realize it is going to involve a whole change in lifestyle. It's hard for me to explain it because I see the way farmers are now—they're part of this whole system. But they're beginning to change, realizing that bigness in farming isn't everything, how important land is, not just owning it but taking care of it.

I want to stay involved. Exactly how, I don't know yet. They are talking about starting this collective in Lowry, so that people can support each other and continue with the work. It has grown in a way we are involved in energy problems all around the country—people contact us from all over the country: people fighting strip-mining, anti-nuclear people, the Indians in the Black Hills. It is a possibility I'll be involved in that. But I have a strong pulling toward living off the land. I just really love working with the soil. The idea of having a few acres, being self-sufficient, producing your own energy. I really believe in the struggle we're in and I can see all these connections with movements across the country—it is people saying, "I want a say in how my life is run instead of having corporations having the say." I would just like to be an example.

The towers are falling one by one, hurrah, hurrah

The towers are falling one by one, hurrah, hurrah

The towers are falling one by one,

It's down with the towers and up with the sun

As we all go marching out—to the fields

—with a wrench—in our hands

Boom, Boom, Boom!

. . .

The towers are falling four by four, hurrah, hurrah

The towers are falling four by four, hurrah, hurrah

The towers are falling four by four,

We're beginning the energy war

As we all go marching out—to the fields

—with a wrench—in our hands

Boom, Boom, Boom!

—from "The Towers Are Falling"
by The Unity Theatre, Minneapolis

8   The Towers Go Up:
The Towers Come Down

Sometime early in the morning of August 2, 1978, a giant steel
tower crashed to the ground on the Ralph Martin farm by the
south shore of Lake Reno, about five miles northwest of Lowry.
Many more towers would follow; the powerline protest had
entered a pathsetting new phase.

In the spring of 1978, company crews had moved swiftly across
Pope and Stearns counties, pouring concrete to anchor tower
bases firmly in the ground. The rest of the towers were then
erected and, by late summer, the powerline was strung in an acro-
batic show of precariously perched linemen and hovering helicopters.

With this impressive display of modern technology, the protest
was supposed to be over. All their legal remedies had been ex-
hausted. The farmers had lost. Many were facing criminal charges
stemming from the previous winter's confrontations with the
state troopers. When energy controversies elsewhere had reached
this point, people had always known when to give up.

The *St. Cloud Times* counseled the farmers to adopt a similarly
realistic attitude when the line was first tested that October:
"The energizing of the controversial high voltage powerline
across Minnesota may be a bitter pill to swallow for those who
have opposed it so long. However, opponents should channel any
bitterness into constructive political activity rather than into
fruitless destruction. The line . . . is now operable. Powerline opponents have lost their battle."

But it was not going to be that easy. Not this time; not this
place; not these people. The real battle had just begun.

GASP

During the confrontations with the state troopers, a new organization had been created at Lowry. It was called GASP, the General Assembly to Stop the Powerline. GASP became the spearhead of the protest. It held regular meetings at Lowry and Elrosa to keep people together, published a weekly newsletter, *Hold That Line*, to spread the word, and established a legal defense fund, to provide assistance to those facing charges.

GASP was a farmers' organization, but it got important help at this crucial time from Cities' supporters. George Crocker and Debbie Pick, a dedicated and capable young woman from Minneapolis, worked full-time in the GASP office, which they had established at Mrs. Flynn's house on a quiet residential street on the south side of Lowry, overlooking the powerline. They spent much of their time putting out the newsletter. *Hold That Line* proved to be an instant success, eagerly awaited in protest households all over western Minnesota. Among its contents were reports on the most recent developments (e.g., "On May 11, the utilities started pouring concrete on Pederson's land, by helicopter"), and what was happening next, and where to be if you wanted to be part of it (e.g., "The Stoen's and Pederson's land is a good place to make a stand using the tactics of non-violent civil disobedience. . . . Because of the increased construction work, the time is right for us to let the people around the state know that we are alive and well. And the troopers won't be in our way until they figure out what is happening . . . come to Lowry Hall at 10:00 A.M. on Wednesday, May 17").

It offered protest philosophy:

The protests have taken many forms over the past several years. During that time every single avenue within the legal system has been tried, and protesters have been lied to, cheated or taken for fools every step of the way. Every step taken in that "legal" labyrinth has brought protesters one step closer to a dead end. . . . There was, and continues to be no satisfaction from this direction.

Instead, satisfaction is coming from the fields. Every time a tower falls, there is a bit of satisfaction. Every time the steel is bent, stakes are pulled, insulators are broken or construction equipment doesn't work the next morning, there is a bit of satisfaction. It is a mistake to label these as acts of vandalism. Vandalism means "ignorant destruction of property," and it would be a big misunderstanding to think that these acts are creatures of ignorance.

It also offered advice. From "a little quiz on how to deal with agents":

sample question:
If an agent comes to my door, I should
1) Invite him in for coffee and cake
2) Tell him I don't know anything or anybody
3) Pick the lint off his suit
4) Tell him I have no legal obligation to talk to him and he should leave the premises immediately.

sample answer:
If you answered (1), you are contributing to our lawyers' migraines and you don't deserve any points at all.
If you answered (2), you're inviting trouble by making a statement they could prove wrong, *0* points.
If you picked (3), and if you keep your mouth shut, since he'll probably leave anyway you got *4* points.
If you answered (4), congratulations. You get *20* points for paying attention and doing the only thing you should.

With the help of Patty Kakac, who drove her pickup to Lowry after her farm chores every day, GASP flourished and *Hold That Line* became the glue that held the protest together. It spread the word about what was happening in western Minnesota to a readership that eventually reached over two thousand people in thirty-three states.

Equally important for the protesters was the legal defense fund. Through donations, raffles, and other appeals, GASP raised tens of thousands of dollars to assist protesters facing prosecution. What is more, a young supporter from the Cities, John Kearney, assembled an outstanding group of criminal lawyers to defend them in court for nominal fees. Included were Ken Tilsen, one of the most respected criminal attorneys in the state, and Mark Wernick, a rising young lawyer dedicated to the farmers' cause. Tilsen and Wernick handled many of the most important trials with a remarkable level of success. By the fall of 1978, they had won acquittal on thirty-four of thirty-six counts that had been decided by juries. The farmers now had the "smart lawyers" on their side and those who would prosecute them knew it.

Late spring is planting time in Minnesota and farmers were out in their fields on their tractors at the crack of dawn. Pope and Stearns county farmers were not pleased to find ugly weeds of steel sprouting when they went out to sow that spring. Those towers could not be ignored. Through *Hold That Line* the call went out to everyone to come to what were called "wiener roasts." After an exhausting day in the fields, thirty to a hundred farmers and members of their families of all ages would gather round a fire and roast wieners and just see what happened to the towers nearby. Occasionally a few people would slip away from the crowd and do some experimenting.

A Vulnerable Link in the Energy Network

One thing that was learned soon was that it was easy to get the bolts out, but quite a trick to bring a tower down. Roberta Walburn, a reporter, was with about sixty protesters at one of the early wiener roasts on an evening that June and wrote about it in the *Minneapolis Tribune.* They were gathered on the Clarence Arceneau farm northwest of Elrosa in Stearns County, where earlier that day workers had begun to assemble two towers. One was about thirty feet high, the other about fifty feet when they had quit for the day. Walburn's article began:

> Around the bonfire, in the middle of the night in the middle of a cornfield, the boy was puzzling over a problem: how to cook a hot dog.
>
> "The end of the stick is as big as the wiener," he said. "How you supposed to put it on?"
>
> Not far away, within sight and hearing distance of the fire, other people were puzzling over another problem: how to bring down two partially assembled powerline towers.
>
> "Oh, it would be so much easier with a stick of dynamite," said one powerline protester as he watched others attacking

the thin steel bars of the towers with wrenches and gloved hands.

The night wore on. The hot dog got cooked. The two powerline towers tumbled down.

Later in the article, she described in more detail what had happened:

Protesters removed bolts on the four legs of each tower. Then they used their hands or body weight or steel bars to tilt the towers a bit.

"This is the easy part," said one unimpressed protester as a tower swayed. "I've seen this before."

The hard part, he said, would be getting the towers to topple, something they'd never done before.

Another protester shook his head as the noise of steel banging against steel rang out in the quiet country night.

"This is a terribly slow method," he said, "and terribly noisy, too. Noisy as hell. There's got to be a better way."

Protesters said they have experimented with acid, hacksaws, torches, and wrenches in unsuccessful attempts to bring towers down in the past.

About this time a pool was started. For a dollar one could buy a ticket and predict when the first full tower would fall. The person with the closest guess would win a third of the pool. Another third would go to the legal defense fund, and one third would start the pool for the next tower.

Another discovery was the vulnerability of the insulators. To keep the current flowing along the powerline, instead of being sidetracked in a 400,000-volt flash down through the tower to the ground, each of the conducting wires is supported by a string of twenty heavy glass insulators. It was found at the wiener roasts that the insulators were relatively delicate objects. Rifle fire or even wrist rockets (sophisticated sling shots) would send them crashing down. As the summer wore on, the ground around the towers became littered with glass.

A third vulnerable element was the aluminum conducting wire of the powerline. An inch-and-a-half in diameter, the wires were evidently not a difficult target for a high-powered rifle with a telescopic sight. A Stearns County farmer was quoted in a newspaper as reporting, "One thing they have been doing in Pope County is shooting the line with a 30.06. If you shoot the line, it leaves it hanging by a few threads and when they energize it, it burns itself out."

Naturally, the power companies were aware of these developments. They announced they were hiring a Chicago-based security firm, the Burns Detective Agency, to provide private guards to patrol the line. It was not an easy task they were asked to perform. In Minnesota alone were 176 miles of powerline with 685 towers, most of them out in farmers' fields and remote from public roads. What is more, they were operating in country that was alien to them, but home to their farmer-adversaries.

At a cost of twenty-six thousand dollars per day a private security force of three hundred guards was deployed. Burns Agency vehicles cruised the country roads at night, and high-speed helicopters buzzed back and forth along the line at all hours. But they were up against an elusive adversary on its home ground. A June issue of Hold That Line likened their quarry to the tiny fish that had halted construction of a Tennessee dam and asked "If a 2-inch snail darter can stop a $116,000,000 project, how big does the darter have to be to stop a $1,246,000,000 project? What if the darter has a nice cornfield to dart around in?" Under these conditions, it is not surprising that the guards were frustrated.

Also not surprising were the apprehensions that the guards and the protesters had about each other, and the tension and hostility that developed between them. Taunts and gestures in encounters on back country roads, sharp exchanges between observers and observees at wiener roasts, and close tailing on drives home from night-time meetings were regular features of their encounters.

To make matters worse, UPA hired another private detective firm from Minneapolis, Centurion of Minnesota. For three weeks in July, Centurion employees tried to strike fear in the hearts of the protesters, and to stir up trouble between them and the powerline workers. At one point they were responsible for an incident of nearly tragic proportions.

The first fatality of the CU project had occurred in July; a powerline worker had fallen to his death in an accident on tower 1180, west of Lowry. A Centurion employee then produced a fake version of the powerline newsletter. It looked like *Hold That Line* and had a similar style; it said, "Thank you to whoever assisted the struggle by tampering with tower 1180. Your work showed that the people will not remain passive."

It was a complete fabrication, but when the fake newsletter was distributed to powerline workers, they were up in arms. This led to an incident the next evening that began with workers forcing a pickup truck carrying Math Woida and John Kooiman, a St. Cloud State University student, off the road and ended with a mob of over sixty workers, some of whom had been drinking and were armed with knives, threatening about twenty-five protesters who had arrived on the scene. Pope County sheriff Emmons found he had no control over the situation and it is remarkable no one was seriously injured. That December, the state board of private detectives suspended Centurion's license for this and other powerline activities.

By the end of July 1978, the power companies were worried. "Bolt weevils" were out nearly every night removing bolts from the tower bases. Stringing of the powerline wire was nearly complete in Pope County, but there was now serious question whether the towers would stay up and the insulators and conducting wire would remain intact. Three hundred guards and local law enforcement personnel had proved unable to quell the sabotage. On July 27, UPA president Jacob Nordberg and CPA president Charles Anderson paid a highly publicized call on Governor Perpich's office to ask for help. As Nordberg put it that day, "We're in real serious trouble and I hope the state understands this." An aide to the governor indicated he would not send out any state troopers, but that the BCA (Bureau of Criminal Apprehension, a state version of the FBI) was "in the vicinity."

Towers Begin to Fall

About that time the towers started to fall. As already reported, the first went down near Lowry on August 2. The same night another tower a few miles to the east, on the Dale Massman farm near Villard, lost its bolts and toppled off its base; remaining upright, it began sinking slowly into the muddy swamp around it. Despite frantic efforts to prop it up, it finally tumbled over two weeks later in heavy winds during a storm. In the meantime, on August 8 and 9, two more towers had come down, both in Stearns County, one near Sauk Centre and the other northwest of Elrosa.

The *Pope County Tribune* reported that in addition to the four towers that had fallen, "up to twenty" had been discovered unbolted. In a telephone interview with the *St. Cloud Times*, one protester stated that an effective new technique for toppling towers had been discovered. All it takes is two people, a set of wrenches, and pressure in the right spot, reporter Maureen McCarthy was told. The discovery was made by a group of protesters gathered west of Sauk Centre early in the morning of August 8. They used the procedure successfully again the following night.

For all the bravado such a phone call represents, toppling a tower must have been a frightening experience. The towers themselves are huge, rising to a height of 150 to 180 feet and weighing several tons. Once they begin to fall they probably hit the ground in two or three seconds. When the line is energized, there is the possibility of electrocution, as an enormous 400,000-volt discharge occurs when the tower strikes the ground. Beyond the physical danger was the greater danger of a long term in prison. There was surely a constant awareness among those involved that taking down towers was a very risky enterprise.

The power companies sought desperately for a solution to their "problem." At their urging, Governor Perpich issued a formal plea for help from the FBI. He sent a letter to attorney general Griffin Bell asking that the FBI launch an investigation. The head of the Minnesota BCA was quoted in the press as saying that the FBI had greater investigative resources than his bureau and, furthermore, anyone apprehended might have a better chance of being convicted in federal court than by local juries. However, the FBI immediately turned down the governor's request, saying it had no authority to enter the case because no federal statute had been violated.

At the same time, newspapers throughout west-central Minnesota began carrying large ads with stark headlines, $50,000 REWARD and POWERLINE VANDALISM HOTLINE: 800-442-3013. With the promise of fifty thousand dollars for information leading to the arrest and conviction of those toppling the towers, people were asked to call any time, day or night. Confidentiality was promised. The bolt weevils now had a price on their heads.

Later, when more towers had fallen, the bounty was raised to a hundred thousand dollars. That was, a GASP press release noted, "except for the Shah of Iran, the largest reward offered anywhere for anything."

In an effort to turn public opinion against the protesters, the power companies also launched a major public relations campaign. The central theme of the campaign was that the "vandalism" was taking money from the pockets of CPA and UPA customers. Power company spokesmen were quoted on every major news outlet in the state saying that the vandalism had already cost CPA and UPA households $139 million. A UPA press release explained,

> Actual vandalism, delay and other costs due to powerline opposition are $43.5 million to date; but since the cooperatives do not have general funds available to pay for these increased project costs, the money must be borrowed. Current interest rates are 8.7%. Over the 34-year payback period, the consumer members of the cooperatives will end up paying about $139 million in interest and principal for the vandalism and delay costs. That amounts to $470 per customer account.

At the same time, the news media began receiving weekly press releases from the companies, detailing the week's "vandalism costs," usually amounting to several hundreds of thousands of dollars.

With the resources available to the companies, this campaign was quite effective. When the line went into commercial operation the following year and CPA and UPA customers began paying off the $1.2 billion in loans for the CU project and their monthly rates went shooting up, there was a widespread perception, spawned by company propaganda, that the increase was primarily due to the vandalism. In fact, the tower topplings and insulator breakage were responsible for only a miniscule fraction, less than 1 percent, of the rate increase. But that was not what most Minnesotans were led to believe.

Protesters had difficulty countering this public relations offensive. They were sharply queried about the $140 million cost of vandalism wherever they spoke around the state. One reason it was difficult to reply was because they were frankly puzzled about where the figure came from. This mystery was not solved until several months later when persistent inquiry by the *Pope County Tribune* produced a breakdown of the figure from UPA. It turned out that the companies had included in their reported "vandalism cost" all of the expenses associated with the local, state, and federal regulatory proceedings and the costs associated with the subsequent changes in routing and construction mandated by the regulatory agencies.

A later study released by the U.S. General Accounting Office in November 1979 breaks down the increased costs of the CU project in a way that clarifies the contribution of "vandalism." On top of the initial 1973 estimate of $537 million for the CU project were $703 million in unanticipated expenses: $215 million extra for the coal mine, $346 million extra for the power plant and $142 million extra for the powerline. According to the GAO study, the extra cost of the powerline broke down as follows:

Regulatory agency mandated siting and construction changes	$28.9 million
Delays due to regulatory process and court actions	23.6
Underestimate of interest payments	29.0
Underestimate of converter substation costs	17.0
Underestimates of spare parts, insurance, taxes, etc.	11.5
Underestimate of right-of-way acquisition costs	25.6
Vandalism and security costs	6.1
Total	$141.7 million

Thus, of the total cost of the CU project, $1.24 billion, "vandalism and security costs" amounted to $6.1 million, or one half of one percent. This did not stop the propaganda campaign. A MAPP brochure circulated in 1980 continued to assert: "The cost for repair and replacement of damaged equipment is already estimated at more than $140 million, figuring principal and interest over the life of the project's financing."

No more towers came down for a while, but the power companies had plenty to worry about. Insulators were breaking every night. In less than a year, the companies reported fifty-five hundred insulators had been shot out. It became a full-time job just to keep replacing them. A new reward of a thousand dollars was offered for information leading to the arrest and conviction of anyone damaging insulators.

In the meantime, electrical testing of the powerline had begun. On September 27, 1978, the last of the 1685 towers was erected. Three weeks later, after one night's barrage brought down 362 insulators and resulted in a last-minute delay, the line was energized. Although the North Dakota power plant was still a year from operation, the Dickinson substation for converting DC to AC near Delano, Minnesota, had been completed and electricity was passed through the line from the Minnesota grid.

Soon thereafter, the Twin Cities' Northern Sun Alliance, a group known principally for its opposition to nuclear power, planned a demonstration at the Dickinson substation to show urban support for the farmers' cause. Gathered under the powerline that cold, blustery Sunday afternoon in late October was a natural coalition: rural, anti-coal and urban, anti-nuclear, united in their commitment to a decentralized energy future. Because they originally advertised it as an "occupation" of the substation, they got a well-organized reception from a large detachment of law enforcement personnel. After two-and-one-half hours of speeches, a delegation of twenty, including one teenager, marched toward the front entrance of the substation, where all were immediately arrested.

Their trial showed the value of a good lawyer. Tilsen told the jury that they were involved in a "political trial," with the issue being whether or not people have the right to assemble and express themselves on important social issues. He was aided by eloquent statements about the powerline by many of the defendants and by several farmers who appeared as witnesses on their behalf. What originally looked to be an open-and-shut case of trespassing ended with all the defendants being acquitted.

GASP's legal defense committee was equally successful in other battles on the legal front. Near the end of September, the special prosecutor in Pope County suddenly announced he was convening a grand jury. Many of the protesters feared he would call in a few people, grant them immunity, and demand that they reveal who was responsible for the sabotage. After consulting the legal defense committee lawyers, *Hold That Line* offered the following advice: "If you are subpoenaed by the Grand Jury, (1) Take it; (2) Call Deborah Pick or John Kearney; (3) Stay calm. We will have lawyers lined up in advance." By the time the grand jury met, the protesters were prepared to meet it with lawyers and a unified front of silence.

In the end, however, all the grand jury did was to hand down indictments that duplicated charges already pending against them from the confrontations the previous winter. Those charges, most of which were in the catch-all category of "obstructing legal process," had hung over their heads as a threat for many months and the grand jury indictments reaffirmed that threat. The special prosecutor never chose to bring them to trial, however, and in February 1979, when Pope County again had its own elected county attorney, he announced that he would move to dismiss all the charges. Sixty-five complaints against fifty protesters arrested in Pope County during the 1978 confrontations were simply dropped a year later.

A New Governor

Meanwhile a fifth tower had fallen. Three days after Christmas, in Traverse County near the North Dakota border, a tower went down in a snowy grove of trees. It was just in time to greet a new governor. Al Quie, himself a farmer, who during his campaign had professed to be disturbed by the way farmers had been treated out by the powerline, took office on January 4, 1979.

No sooner had he sat down at his desk that day than he received a telegram from the protesters. The subject was health and safety. The telegram stated:

> After limited testing of this line, many people have experienced disagreeable and intolerable effects. For example, many people have experienced headaches as well as skin rashes. . . . Immediate action is required and this is our proposal: That there be an immediate halt to testing of this line until there has been thorough investigation. . . . As an act of good faith insuring the enforcement of a moratorium on test-

ing, we stipulate that the downed tower in Traverse Co. not
be replaced. We are confident you will use the powers of your
office to work toward a solution to this problem. . . .

However, negotiations with the new governor did not go well.
Invited to come out to Lowry to meet with the folks, he thought
about it for a while, but eventually declined. Another tower was
unbolted and a siege of insulator disease set in, with over 350
insulators per week being shot out during January and February.
Quie found himself under pressure from the power companies
and from the REA in Washington to crack down on the protesters.
In a televised press conference, agriculture secretary Robert Bergland called on farmers to turn in their neighbors if that was what it
took to quell the vandalism. The BCA was out in force, with *Hold
That Line* reporting that "security is so tight you can't breathe."

On March 21, Jacqueline Thurk, thirty-one, a farmer from Villard, who along with her husband Kenny had long been active in
the protest, was convicted by a Douglas County jury of assaulting
a state trooper. She was accused of hitting him with a picket sign
that read WHAT ABOUT THE NEXT GENERATION? during the powerline demonstration at the Alexandria cement plant in February
1978. Proclaiming her innocence, she had turned down three offers to plea bargain. The conviction of this churchgoing mother of
four, an upstanding member of her community, whom even the
prosecuting attorney admitted was "quite a marvelous person,"
upset the protesters. They had come to expect to win in court if
judged by a jury of their peers. Although the charge had been reduced from a felony to a misdemeanor and the sentence was suspended by the judge, it was nonetheless a shock.

The day after the jury returned its verdict in the Thurk trial, the
legs were sawed on a corner tower in Pope County and it came
crashing to the ground. That was number six.

Two more trials remained. The first involved six men arrested
in connection with the incident back in March 1978, when the
window of a truck had been shot out, showering a powerline
guard inside with glass and bullet fragments. Fortunately, he had
not been hurt seriously. The six charged included four farmers
with land along the powerline, Harold Fischer, Darrell Bartos,
Tony Bartos, and Math Woida, and two outsiders, William T.
Hansen and his son, both farmers from Wadena, some fifty miles

to the north. They had been turned in by Fischer's wife, who reported she had heard the six men planning a trip to the powerline construction yard where the shooting occurred.

Only Hansen went to trial. By the time the prosecution was prepared, none of the defendants or their wives would testify, even if granted immunity. The prosecution was stymied; it had no case. At the last moment, however, the judge ruled that Hansen must have threatened Fischer and his wife to keep them from testifying and therefore statements they had given to the BCA, previously ruled inadmissible, could be introduced as evidence. Hansen was convicted on two felony counts and sentenced to two years in prison; he is appealing that conviction. With no one willing to testify against them, the charges against the others were later dropped.

Four days after the Hansen jury delivered its verdict, tower number seven came down east of Elrosa, near John and Alice Tripp's farm.

The final trial was another example of Ken Tilsen's courtroom brilliance. Two Pope County farmers, brothers Bob and Dean Oeltjen, were accused of holding two powerline guards at gunpoint in a night-time confrontation at their farm. Tilsen maintained that the Oeltjens had not pointed the guns at the guards and that they were actually defending themselves and their families against harassment and holding the men for the sheriff. They were acquitted on all counts.

From 1976 to 1978, 120 people were arrested in connection with the powerline protest and charged with approximately two hundred separate criminal counts. Many had serious charges hanging over them for long periods of time. But in the end, 75 never went to court; the charges against them were dropped. Fourteen more pleaded guilty, all but one to misdemeanor or petty misdemeanor charges. Thirty-one did go to trial; 27 were acquitted and 4 were convicted, 1 on felony charges. For quite a while it seemed possible that many lives would be broken with long prison terms. Thanks to the legal defense fund and the GASP lawyers, in the end only three were sentenced to jail, with terms ranging from five days to two years.

On August 1, 1979, the CU powerline officially went into commercial operation, carrying power from the North Dakota power plant into Minnesota. However, on August 24, a tower toppled in Stearns County and three days after that another fell in Wright County. On November 5, the tenth tower came down, Meeker County's first. The following March, Stevens County out west found it had bolt weevils, too. A few days later, tower number twelve came down in Kandiohi County near the eastern end of the line.

By the summer of 1980, state officials and the power companies were practically begging the FBI to come to their aid. In a major speech, Minnesota attorney general Warren Spannaus announced he was sending an aide to the Justice Department in Washington to "impress upon the officials there the seriousness and urgency of our demand [for FBI assistance]." At the same time, efforts were underway in Congress to pass a law making taking down towers a federal crime. After a flurry of gunfire had shattered enough insulators to put the line out of operation for two successive days in July, the REA announced in Washington that it was seriously considering taking over ownership of the CU powerline, if that was required to bring the FBI in.

As this book goes to press, the long-term reliability of the CU line is still in doubt. Towers thirteen and fourteen came down in early September and the following week, on September 9, 1980, UPA and CPA formally requested that the REA assume ownership of the powerline. REA's Frank Bennett, now head of the Power Supply Division, indicated that the request would be approved and noted, "REA ownership makes these acts of terrorism a federal offense." FBI intervention—federal agents and a federal grand jury —appears inevitable. Some people may go to jail, but the farmers seem determined to continue the fight.

Vandalism or Sabotage?

Taking down towers is a dangerous, desperate act of rebellion. Only a few short years before, almost all the protesters would have abhorred such "vandalism." Now they applaud and call it "sabotage." Their experience with the powerline has been an education. Far better than any secondhand account, their own words express how they have changed and how they feel about the towers coming down:

Patty Kakac is a kind, gentle, softspoken young woman who lives on a farm in Douglas County several miles from the powerline. "When I hear about a tower going down I rejoice and rejoice again. I report it to all the neighbors as good news. You know it is really a hard thing to explain why you feel that way about this thing. You are raised on a farm very strongly to respect property. You have to, otherwise you couldn't exist. You couldn't have all your machinery, all your equipment, all your cattle, your animals; everything is worth money, and people could come and take it any time they wanted. So you learn that; it is instilled in you. And yet, something in us just rejoices every time that powerline tower goes down and I know the neighbors feel the same way.

"It is a hard thing to explain. I guess it is because of what the powerline is, what it means. I guess the people in the country are so afraid of some power coming and taking over them. It has been happening slowly over the years. They have been shoved into a rut of being dominated and controlled by oil companies and chemical companies. In these things, the control was gradual, but the powerline was all of a sudden such a huge project forced on them. . . .

"It's funny. Several years ago I would have thought pulling towers down was extreme. Not now. I am almost to the point where it is not extreme enough.

"It is real sad. I wish it wouldn't take this. But people wouldn't pay attention to the health problems or the way the powerline was put across; they wouldn't pay attention to the struggle that the farmer goes through until something drastic brings their attention to it. As soon as a tower goes down, almost the whole state knows about it. Maybe it puts a question in people's minds —What is going on? What is happening here?"

Anne and Mike Fuchs began farming in 1938. They are now retired, but their children continue to work the land in western Stearns County.

Anne Fuchs: "When the towers go down—I don't know who does it and I don't want to know—but I don't think it is meanness. It is just we never was heard. If a lot of people only knew how we were treated. We talked, and we knew we had talked to the wind. Now a lot of people think that the only thing we have left is if a tower goes down—we have to do something to make them listen."

Mike Fuchs: "There are a lot of people who don't say nothing, they don't go to meetings or nothing, but when they talk to us they say, gee, isn't it about time that another tower goes down, just as if we should do that."

Gene Quinn is in his mid-fifties. His family has lived on his land for several generations. He is a very successful farmer, with five hundred acres in Meeker County. He is known as a moderate man, highly respected in his community. "Do I approve of it? Do you approve of the Revolutionary War! I just feel the rights I had were sorely abused. It was an insult, was what it was, just a plain insult. As an individual, they had no respect for you. They don't care what they do to the people or the land.

"I don't feel the process was carried out in a legitimate fashion in the first place. To me it was apparent the utilities were ignoring the process whenever it didn't meet their aims. I don't feel the power companies or the state gave the people a chance.

"I really think people sense guerrilla warfare as a way of making themselves heard even if the powers that be would ignore them."

Dick Hanson is a college student, a farmer, and a political activist from Glenwood in Pope County. In June 1980 he was elected as one of four Minnesota representatives to the Democratic National Committee. "We have been through the courts, through the legislature, through the executive, and we have been totally ignored. All the so-called legitimate channels broke down, and it left no other alternative, except to turn over and play dead. If we really cared about our families and our livelihoods as farmers out here, we had to do something. We had to take it into our own hands. It is self-defense, I think.

"Many different people support the sabotage and I think it is because of the rough way the power companies along with the government rammed this thing down our throats. They had the courts working for them, they had the legislature, they had the money and power to push all these things through. Now we have the power. They cannot put a security guard behind every corn stalk and every tree and under every tower twenty-four hours a day. It certainly is unusual, but I am not so sure it is out of the ordinary. Take the Boston Tea Party incident or a heck of a lot of other incidents where people had to take things into their own hands because there was no legitimate avenue open. That is the case now."

Minnesota's Long-Range Energy Plan

The approach to the new governor did open up one avenue of communication that for a while looked quite promising. In March 1979 Quie sent his aide Robert Stevenson out to Lowry to meet with the protesters. About 150 crowded in the Barrel Inn, a bar and grill on Lowry's main street, which had become the meeting place for the protesters after the Lowry town fathers had decided they could no longer use the town hall. Stevenson made it clear at the outset that Governor Quie had decided not to follow in the footsteps of his predecessor and become personally involved in the powerline dispute. He seemed receptive, however, to the farmers' proposal that the governor sponsor a public forum on powerline issues. This was the beginning of several months of negotiations about a forum.

On reflection, the governor soon decided that he would rather have his public forum consider energy in general instead of the powerline in particular. In early April, Stevenson wrote to GASP, "The open forum concept is a good one for discussing energy issues. The Governor would like me, as his representative, to explore the idea further. . . . However, this open forum idea is not useful to solve your problems with the existing CU powerline." He closed with a plea: "The continuing violence and property destruction is currently the major obstacle to communications. I urge your members to discourage those few vandals who are aggravating this situation, and lessening public sympathy and understanding for your problems."

Stevenson then invited representatives of GASP to meet with him and state officials at the governor's office in the state capitol. To make sure the focus would be broader than the powerline, he also invited representatives of a state-wide coalition of environmentalists, CO-REG (Coalition of Rural Environmental Groups), and representatives of a consortium of electric utilities, the Minnesota/Wisconsin Suppliers Group, to participate in the negotiations.

Out of these negotiations came a proposal, initiated by the farmers, for a "Minnesota Electrical Energy Policy Forum," a one-day symposium in St. Cloud. After brief introductory remarks by Governor Quie, the program would begin with a presentation by the director of the Energy Agency of Minnesota's comprehensive, long-range plan for electrical energy development. Two critiques of the plan would follow, one by the farmers, the other by the utilities. Amory Lovins, noted author of *Soft Energy Paths*, was contacted and he agreed to speak for the farmers, provided the forum could be scheduled in November to coincide with a visit he was already planning to Minnesota. The utilities talked about having Donald McCarthy, president of Northern States Power Company, present their critique. To allow time for thoughtful preparation of the critiques, the Energy Agency was to release a detailed paper describing its comprehensive long-range plan, about two weeks before the forum.

There was agreement on this proposal by the farmers and the utilities. Arthur Sidner, head of the EQB and the State Planning Agency, who chaired the negotiations, was also enthusiastic. The only people present who did not show enthusiasm for the proposal were staffers from the Energy Agency. They indicated they were not at all sure their agency could be ready with a plan by November. This prompted an angry response from the farmers. Dr. Wendell Bradley, a farmer and a physics professor at Gustavus Adolphus College and a leading member of CO-REG, said in a statement to the press after the meetings, "The Energy Agency people were the only ones expressing reservations about this forum. . . . Since they have been issuing certificates of need for some years now, it is outrageous that they are not ready to explain to the public the basis for their decisions." He added that if the agency did not have a comprehensive long-range plan, it was about time they began a public process to develop one and the forum proposal was precisely what was needed to begin that process: "We are very pleased that this process will finally take place; it's about time the Energy Agency involved the public in developing a plan for electrical energy development instead of just rubber stamping utility proposals. As is well illustrated by the CU powerline, in the past the agency has gone through the motions of deciding on a certificate of need for each proposed utility project, but since it never had a comprehensive plan of its own, the certificate of need process has been a charade. We hope this forum will begin a public debate of the critical energy choices we face and result in a long-range plan that comes from all the people of Minnesota, not just the utilities."

By pressing the Energy Agency to present a plan at a public forum, the farmers were after much more than something to talk about one day in St. Cloud. They were trying to force a change in the rules of the energy game in such a way that they and other members of the public might really have an opportunity to participate in the formulation of energy policy. They *knew* the Energy Agency did not have a comprehensive, long-range plan for Minnesota's energy future. As in most states, energy policy planning in Minnesota has been done piecemeal, in a wide variety of decision-making forums. For example, comprehensive planning for electrical energy development is largely in the hands of the utilities, through such regional planning organizations as MAPP. In leaving the comprehensive planning for energy to the utilities, the state, in effect, allows those interests to decide the boundaries of "realistic" energy options. These boundaries become the shared assumptions of the fraternity of state and utility officials. But once the boundaries are established, the most the state can do is tinker a bit with utility plans. There is no way of challenging the basic assumptions underlying energy policy and possibly charting a radically different energy course.

If the Energy Agency had to present its proposal for a comprehensive energy plan in this *ad hoc* forum, there would at last be a focus for public discussion and debate of the state's energy future. The assumptions and values underlying the proposed plan could be understood; alternative assumptions and values could be considered and alternative plans could be proposed.

In this sense, the energy forum proposal, which grew out of the powerline dispute, would have been an innovative, important policy-making instrument. But it never took place. At a final climactic meeting in September the whole thing fell apart. The meeting began with the utilities announcing that, on reflection, they had serious reservations. The director of the Energy Agency, Algernon Johnson, came for the first time, listened for a while, and then announced firmly that the agency had no plan to present.

A few days later, EQB head Sidner, who had spent months on the negotiations, sheepishly announced that the whole thing was off.

Nothing to Worry About or Guinea Pigs?

In the ensuing months, Sidner also became deeply involved in dealings with the farmers on another front. Ever since the line had been energized in 1978, there had been reports by farmers living nearby of an abnormal incidence of certain symptoms, notably headaches, nosebleeds, rashes, respiratory problems, dizziness, and fatigue, when the line was on. (It is easy to tell when current is flowing through the line—it emits a very distinctive noise, which sounds a lot like bacon frying on a stove.) There were also reports of problems with livestock, notably fewer pregnancies, more abortions, increased stress, and decreased milk production.

On June 25, 1979, congressman Richard Nolan sponsored a public hearing in Sauk Centre and invited anyone who felt they were experiencing adverse health effects to come and testify. One of Nolan's aides presided over the meeting and up on the stage with him to receive testimony was George Pettersen, the state commissioner of health. For several hours, they listened to anyone who wanted to come forward and speak. Although purely anecdotal, the reports were remarkably consistent. Similar symptoms were described by farmers all along the line.

At the meeting it was pointed out that the pattern of symptoms described was similar to that associated in the scientific literature with air ion effects. It was air ions that Robert Banks, who headed the 1977 health department study, had pinpointed as a possible DC powerline health hazard deserving of immediate investigation. Among the most reputable experiments with air ion health effects were those of A. D. Krueger and collaborators at the University of California at Berkeley. Krueger had noted significant adverse health effects from positive ions in concentrations of a hundred thousand ions per cubic centimenter, roughly that found in the vicinity of a +/−400 kv DC powerline. According to Krueger the probable mechanism for the effects was the "seratonin irritation syndrome," in which positive ions cause secretion of seratonin, a powerful neural hormone, which in turn produces symptoms that include migraine headaches, nausea, amblyopia, irritability, hyperperistalsis, rashes, conjunctivitis, and respiratory congestion.

Of course the combination of anecdotal testimony about symptoms and a possible mechanism that might account for these symptoms, while suggestive, is far from scientific proof. Whether the symptoms were in fact related to the powerline was quite uncertain. An important policy question is what to do in the presence of such scientific uncertainty. Should the powerline be left on, or should it be turned off until the uncertainty is resolved? The farmers, who felt they were being used as "guinea pigs," and the power companies, who had much invested in the project, naturally differed in their answers to that question.

There was much talk about "guinea pigs." From the farmers' perspective, the 1977 health department study had pinpointed air ions as a possible health hazard, symptoms similar to air ion effects were being experienced, the health department was planning to use a hundred thousand dollars allocated by the legislature to study, among other things, air ion concentrations in the vicinity of the powerline, and the power companies had refused to accept absolute liability if unexpected health problems did appear. It is perhaps not surprising that they felt they were being used as human guinea pigs.

Much anguish and much activity in the subsequent months centered on the health issue. In September, GASP organized a press tour along the powerline to give the media an opportunity to talk with the farmers about the problems they were experiencing. The commissioner of public safety offered to sponsor a scientist to come out and study the problem. At its September 18 meeting, GASP issued the following reply:

> With the present state of scientific knowledge of the health effects of high voltage transmission lines, as long as the line is energized people are being used as guinea pigs. Even the Minnesota Health Department Study indicated areas of serious uncertainties in regard to health effects. The power companies' refusal to accept absolute liability shows they are not sure either. Numerous medical complaints since the line has been energized cannot be swept under the rug. We refuse to be guinea pigs. We demand a moratorium on the operation of the powerline until these uncertainties are resolved.

There followed a series of angry confrontations, one at a regular meeting of the Environmental Quality Board, which was disrupted by powerline protesters who said they would keep coming back until they got answers to the health questions; another demonstration took place at the health department in Minneapolis. Health commissioner Pettersen's position was that it was unclear what should be done. He wrote to representative Dave Fjoslien in December that he would either "convene a panel of scientists" or "consult with recognized scientific experts."

The protesters responded they didn't want "experts" to decide; GASP issued a statement calling instead for a "powerline jury of peers," twenty-five rural citizens selected at random from the voting roles to review the scientific literature, hear from expert witnesses, and review the evidence of health problems along the line. The jury would function like a grand jury, deciding whether or not there was probable cause to believe (not proof) that the powerline was responsible for adverse health effects. If probable cause were established, the burden of proof would then shift to the power companies. The line would shut down until they could demonstrate that it was safe.

Presented with this proposal, Commissioner Pettersen said he had no authority to institute such a proceeding. According to *Hold That Line,*

> The attitude of the Commissioner can pretty well be summed up by two exchanges. . . . When asked, "How will you know when there is enough evidence for you to act?," he replied, "I don't know." When asked, "If you had *your* choice between a process which studies the effects of this powerline, thereby using people as guinea pigs, or on the other hand a process which would establish whether or not there was reason to believe that the powerline is hurting people, shut down the line if there is that kind of probable cause, and then do scientifically controlled experiments in the laboratory, which would you choose?" He replied, after there was a silence in the room for fully 20 seconds, "I don't know."

At this point, Governor Quie took the health and safety ball from Commissioner Pettersen and handed it to EQB chairman Arthur Sidner. Sidner was asked to lead a "fact-finding mission."

He pledged to do a systematic survey of residents along the power-line to find if they were experiencing adverse health effects. At the same time he hired a Berkeley consulting group that included air ion expert Dr. A. P. Krueger to prepare an up-dated report on what was known about powerline health effects.

In June 1980, Sidner announced the results of his survey of residents. He emphasized that he had decided not to conduct a "scientific" survey, with a group living along the powerline compared to a control group from the general public. Instead, he limited himself to polling landowners along the powerline about their perceptions of problems. Almost four hundred landowners responded to the query, "Have you suffered any adverse health effects which you believe are a result of living close to the power-line?," and to a similar question about problems with their livestock. In addition, they were asked if they had experienced sixteen specific symptoms "which you believe have been caused by the powerline."

Thirty-five percent said *yes* they had experienced adverse health effects which they believed were caused by the powerline; 47 percent said *no* and 18 percent said they *did not know*. (According to the persons who administered the poll, in most cases "do not know" meant they had experienced health problems, but did not know for sure whether or not they were caused by the powerline.) The principal symptoms cited were headaches (29 percent), respiratory problems (17 percent), fatigue (16 percent), tingling (12 percent), stress (12 percent), dizziness (11 percent), nosebleeds (11 percent), rashes (11 percent), eye problems (8 percent), and nausea (6 percent).

These results are clearly subject to different interpretation. The *Minneapolis Tribune* headlined its story, 47% IN SURVEY BELIEVE POWERLINE HAS NOT HURT THEIR HEALTH. On the other hand a substantial fraction of the respondents, at least 35 percent—probably more nearly half—thought they might be experiencing adverse health effects from the powerline. In the absence of a control group from the general population, it was difficult to know whether the incidence of symptoms reported was significant or not.

At the meeting where the survey results were made public, Sidner admitted their ambiguity and said that while they were suggestive of possible health hazards, they did not constitute sufficient grounds for him to recommend shutting down the powerline. They were landowners' perceptions, he emphasized, and they would not stand up as evidence in court. At the same time, he expressed his sympathy for the farmers living along the powerline, saying that, in his view, given the uncertainties associated with health effects, they were indeed being used as guinea pigs. Sidner noted that the powerline was not unlike many other new technologies, in being deployed before their health effects are well understood. He later expressed concern that it could become the 1980s equivalent of Thalidomide, the drug that caused birth defects in the 1960s.

To the farmers, Arthur Sidner seemed a breed apart from most public officials they had encountered during the powerline years. He seemed genuinely concerned about giving the go-ahead for their being used as guinea pigs. There was much speculation about how long he would last in government. As this book goes to press, Sidner's fact-finding mission is continuing.

Co-op Members Have a DREAM

One consequence of the powerline dispute and the dramatic increase in electric rates that customers of UPA and CPA power experienced in 1978 was a realization by many farmers that over the years they had lost control of their co-ops. In reaction, several reform efforts were started.

A prime example is DREAM, Determined Runestone Electric Association Members, organized in 1978 by an outspoken, articulate, and politically shrewd farm wife from western Pope County, Nancy Barsness. In early 1978, she had covered the confrontations around Lowry as a reporter for several local newspapers and radio stations. A member of the Alexandria-based Runestone co-op, she wanted it to go on record in favor of a moratorium on construction of the powerline. When she called Runestone general manager, Joseph Perino, however, she was rebuffed. The co-op board of directors made such decisions, not the members, she was told, and their meetings were held in private. When she asked to see the minutes of the board meetings, Perino replied, "The board minutes are not open to the public. We're not a public organization, Nancy. Let's get that straight."

Barsness realized what had happened to the original concept of democratically controlled co-ops: "Today Minnesota rural electric co-ops are 'cooperative' in name only, with management denying members' involvement in co-op decision-making. To give the impression that members are still involved within the cooperative structure, management offers prizes, free lunches, and entertainment to attract members to their annual meeting. Although attendance is encouraged, participation is not."

Her response was to organize DREAM. The Runestone board soon found itself dealing with angry, informed co-op members, not the apathetic group to which they were accustomed. Within a year, the board meetings and minutes were opened to all members, two directors and the Runestone attorney resigned, and general manager Perino himself chose early retirement.

There were uprisings in other co-ops as well. Customers of Dakota Electric Association, a CPA-supplied co-op south of the Twin Cities, were surprised to find their rates had increased by a shocking 50 percent when they opened their August 1979 bills. A hastily organized meeting filled a high school gymnasium in suburban Burnsville. Thirty GASP members drove down for the occasion and Nancy Barsness spoke. The audience was angry and addressed pointed questions to the co-op officials present. When pressed, CPA spokesman Robert Sheldon was forced to admit that the sharp rate increase they had experienced had virtually nothing to do with the "vandalism"; it was due instead to the fact that co-op customers had begun to pay off the loans for the CU project. Comparably large further increases could be expected in the near future.

The members were up in arms. This prompted the Dakota Electric vice-president to write a letter to CPA president Charles Anderson in November 1979, which concluded: "Disenchantment is a mild word for the resulting reaction in Dakota County. Finally, Charlie, we believe that the present CPA Board of Directors and the Executive Board in particular, are being manipulated by the CPA staff. We think it's time you take control on behalf of the member consumers before it's too late."

Following the meeting, at Barsness' suggestion, Dakota Electric members began a petition drive calling for a full vote of the membership to request that their rates be regulated by the state Public Utilities Commission. The co-ops *had* been regulated until 1978, when a bill had quietly passed the legislature exempting them. This was quite convenient for CPA and UPA, in light of the sudden rate rises anticipated the following year when the CU project went on line. The results of this petition drive and a similar one at another local co-op are still in doubt. Although the necessary number of signatures was obtained, the petitions are being challenged by co-op officials on grounds that some women had signed for their households and only the men are really co-op members.

The reform movement is now spreading to other cooperatives. Among the proposed reforms are Public Utilities Commission regulation of CPA and UPA and adoption of a members' bill of rights, including open board meetings, open books and records, and direct election by local co-op members of their representatives to the CPA and UPA boards.

Energy Policy and Rural Americans on a Collision Course

The story of this particular powerline is not yet over. Nevertheless it is possible at this point to look back and reflect on what it means.

According to many people in the electric power business, the powerline protest was merely a result of a poor public relations job at the beginning. With better selling efforts, such conflicts can be avoided with future powerlines.

One proponent of this view, which might be termed the *public relations* response, is Ted Lennick, general manager of CPA. Asked if he were able to do the CU powerline over again, what changes he would make, Lennick replied, "If we did this whole project over again, I would like the time to go out and talk to people. I think that's basically the only change. I think the project is very sound. I think that by and large it's going to be looked at as one of the good projects that was built for rural cooperatives. I do think that, and again it was because of the situation at the time, I think that we didn't get an opportunity to speak to the people like we should have. Some of that is our problem; some of it was forced on us. I think our biggest effort, if we were to do it again, would be to get involvement by the local people."

Michael Murphy has twice been a leading candidate for Energy Agency director and is currently director of the Upper Midwest Council, an "establishment" research and planning organization in Minnesota. His comments to television reporter Andy Driscoll in connection with a TV documentary on the powerline suggest the level of sophistication among planners who are thinking about the kind of public relations effort that will be required: "It is the responsibility of the utility industry to go out and invest the up-front dollars in what I described in a report that we did a few years ago on power plant siting and coal transportation and things; do what we call exploratory and environmental scanning. And that's not the physical environment; that's the social, human environment out there. I think it's incumbent upon the utility to look at its service area, to closely monitor through its contacts and its people out there and stirring the soup and listening a lot and talking very little to find out what's going on in the world and where people's priorities are, where they're happy, where they're unhappy, how do they relate to the prospects of new power plants down the road somewhere, and the feelings about these things. Identify the problem areas and try to overcome them before you get to the regulatory process, because the regulatory process is not designed to do that. It's very locked in once you get there. Certain irreversible decisions have been made.

"I think that once you go out and do the kind of investigation or environmental or exploratory scanning that I talked about —sure, it's expensive and it's difficult to do—but it's a one-time thing in a way. You have a big initial investment and after that you have a maintenance and improvement process . . . it can even be computerized, for that matter, but it doesn't have to be. It can be a little more human that that, I hope.

"We're always going to have people, be they farmers or urban folks, being against something, you know. The trick is not to just sit down and say, 'Well, you're against and he's for, now let's work it out.' The trick is to try to anticipate those things, to monitor it and learn something about changing trends, so that when you get to a point that you have to deal with a difficult decision or conflict, you know how to work with the folks."

The next level of response is that something more than just better public relations wil be required to get future powerlines through, but some modification of operating procedures and increased allocations for landowner compensation will suffice. No major change in the traditional way of doing energy business will be necessary. This might be termed the *technical fix* response.

Examples of technical fixes frequently mentioned range from routing powerlines underground or along existing highway and railroad rights-of-way to massive purchases of land to create "transmission corridors" that cut huge swaths across the countryside exclusively for powerlines, pipelines, and so forth.

Exponents of this approach include Phil Martin, general manager of UPA, who has stated, "We will have to go through swamps, wildlife areas and along highways in the future. The state agencies who make this extremely difficult will have to change their policies." Another proponent of the technical fix is Minnesota governor Al Quie. A member of Congress during the confrontations of 1978, Quie's response was to sponsor legislation making it easier to use highway rights-of-way for powerlines in the future.

Our view is that neither the public relations approach nor the technical fix approach will prove workable. In this view, the Minnesota powerline struggle is not an aberrant episode that need not be repeated if proper social management techniques are employed. It is instead a harbinger of events to come, a consequence of historical factors that have come together to produce the ingredients for an historic confrontation. U.S. energy policies and rural Americans are on a collision course.

In Washington, the main elements of a national energy policy are emerging. The common wisdom that guides it is briefly this: oil imports must be restrained. Nuclear is, at least for the time being, out of favor. Fusion will not contribute in the foreseeable future. Conservation and solar are question marks. Coal is the answer. The United States is embarking on a program that has as its centerpiece a massive investment in coal. As energy shortages develop, the resulting sense of urgency is likely to bring support for measures like an Energy Mobilization Board to cut "red tape" and neutralize resistance that impedes energy development.

In rural America, the main elements of the resistance are beginning to stir. Farmers and other rural citizens—ranchers, Native Americans, small-town dwellers—are coming to realize that their way of life is threatened.

The threat arises on many fronts. Frequently, as in the case of interest rates and agricultural policies, for example, the threat is apparent, but those responsible are remote and inaccessible; as a consequence there is no obvious way to fight back. Energy projects—strip mines, power plants, or powerlines—are different. Here the threat is tangible, the adversary is identifiable, and direct resistance is possible. Projects like the cu powerline can become symbols of the larger threat to their existence that rural citizens are facing in this era.

This is clear from the way so many farmers in western Minnesota relate their experience with the powerline to the more general perception that they are losing control over the way they live.

For example, Dean Arcenau, twenty-two, grew up on a 230-acre farm northwest of Elrosa in Stearns County. A stalwart in the powerline resistance, he wants to go into farming if he can. In explaining why he wants to farm, he noted, "You do have a little more control over your life. I worked building irrigation systems in a metal shop and I don't like to work under other people. I don't like to be told I can only work eight hours a day if I want to work nine. I moved to the Cities after high school and I just didn't like it. I like the open spaces, the greater amount of freedom we have. I want the peace and quiet and the freedom." From this perspective, the powerline was important: "I know my dad and a lot of neighbors, when chemicals and herbicides first came out, really had a lot of problems and hesitation starting to use something that would alter the way of life that much and to bigger machinery. But they used the economic situation to force you into doing it. You are controlled, but it is not obvious. It was never that obvious that somebody had this amount of power over you—that they could just tell you, 'You are not going to do this; you are not going to have anything to say about it'—until they said, 'We are going to put that powerline here.' "

Anne Fuchs, two generations older than Arcenau, echoed this theme: "Like my son Werner, he says, 'My freedom is taken away.' They just went right over his cornfield; they didn't mind where they went. They just didn't care. You don't have much rights left any more if someone can do that to you. I mean just to give you a feeling of how it felt, I remember when the surveying started there, he didn't say much—he is more or less on the quiet side—but I remember him saying when they come in there and rammed over his land, he said, 'That was the saddest day of my life.' You know the feeling: they can do anything and what can I do?"

Patty Kakac expressed the depth of feeling in her community. "The powerline represents control, something else in control of our lives. Farmers are being pushed in a corner by these things that are controlling them . . . people don't know how to fight back. They can go to their party caucuses; they can vote; but it is just all so frustrating. They can go to meeting after meeting after meeting. . . . They just get tired out trying to find some way to change the situation and nothing is accomplished. The same old thing.

"But the powerline is something people can focus their fight against. . . . After going through all these processes—the courts, the political candidates, the legislature, and the governor—and being frustrated, they still have something that they can do that most people can't do—they can go out and take down a tower. If they aren't capable of taking down a tower, somebody is. And the ones that don't do it, there is something in them that wishes they could do something to fight this control over us.

"The economic treatment of farmers plays a big part in the whole thing. Hardly a day goes by that a neighbor doesn't come by and talk with us about this. And they all talk about the same thing. For years they have been working hard and taking care of the land and they work every day of the year. They don't get holidays; they don't get vacations. But they love that style of living. That is their life and it is the kind of work that you can't explain in words what it means to you. You just feel so close to the land.

"But farmers have gone along with farm economists and loan companies and bankers who have told them over and over again that they have got to get more efficient—the problem is they aren't efficient enough. They have done everything they have been told to do—gotten bigger, used more fertilizer, gotten bigger crop yields per acre, put in irrigation. They are going deeper and deeper in debt, and so far they have been able to borrow more and more money because the value of the land has inflated. But every year their income shows a loss. There are farmers in our area that are going bankrupt and have to sell out. They just don't know what to do any more.

"We have been feeling this control over our lives coming. I suppose looking back I can see kind of a build-up. Roads going here, development going there, all this taking over and destroying farmland and taking control away from the farmers. You are losing control of what you do—of what you plant, what you can haul out of your fields, what kind of animals to have, and dependent on loan companies with the money, who tell you how to plant, what to use on crops, how to sell, when to sell. And the chemical companies—you have to use this chemical if you want to get a crop.

"In order to care for the land you have to be the kind of person who is independent; you have to be able to understand your surroundings and your environment. And only you can understand that. Some of the farmers haven't been doing a good job; I'll admit that. But in order to take care of the land, you have to have an understanding of it and you have to live there and it takes years . . . you get to know the soil, the trees, what kind of water is there. This is all in jeopardy because somebody else comes and controls the land and yet they don't understand how it works and the people who live there can no longer make a living there. It is not just a job you lose. For a farmer it is your home, your job, your lifestyle that is lost."

Such sentiments, echoed across rural America, portend serious problems for energy policies popular in Washington these days. The traditional path of highly centralized energy systems, with a heavy new emphasis on our nation's abundant coal resources, may not prove to be the path of least resistance in energy.

An historic energy debate is underway in our country. Physicist Amory Lovins has vividly described the energy choice we face as between a "hard energy path," with a relatively small number of large energy sources under highly centralized control, and a "soft energy path," with the emphasis on conservation and many decentralized energy sources under local control. So far the fight has been pretty much confined to the intellectual arena. Many scholarly studies have been performed, demonstrating that a variety of kinds of energy futures are possible, ranging from the very hard to the very soft.

In the political, bureaucratic, and corporate arenas, where the decisions are actually made, so far the fight is no contest. Although there have been some setbacks, as on the nuclear front, there continues to be tremendous inertia propelling our nation down the traditional, hard energy path. That is the way our energy institutions are structured to take us; that is the way the organized energy interests want us to go.

The CU project illustrates the magnitude of the institutional and economic inertia that would have to be overcome to redirect our energy course. In effect it represents a choice to go down the hard path, made for us by our government. To build the Falkirk mine, the Coal Creek generating station, and the CU powerline, the Rural Electrification Administration in Washington arranged for UPA and CPA to receive $1.2 billion in long-term, low-interest loans. But it is UPA and CPA customers, three hundred thousand households in all, who will pay for the project. Each UPA and CPA household has, in effect, taken out roughly a $4000 involuntary loan through the auspices of the U.S. government.

But just consider what they might have bought with a $4000 energy investment instead of electricity from coal. Recent studies suggest that by spending $1500 to $2000 for insulation and sophisticated leak-plugging, most American households could cut their heating and cooling bills in half. The remaining $2000 to $2500 could be used to purchase renewable energy technology— for example a wood-burning stove or a solar water heater.

The institutional and economic inertia behind the choice that was made is enormous. Central to the decision that each UPA and CPA household would take out a $4000 energy loan and use it for electricity from coal was the existence of a bureaucracy in Washington, the Rural Electrification Administration, dedicated to financing electric power projects. Supporting the decision were several institutions and interests, including UPA and CPA, North American Coal Company, and other utilities in the MAPP electric grid. Given the mission of REA, there is no way that the $4000 energy loan could have been used for anything but electric power.

It is enlightening to reflect on what it would take to change even this one small part of our energy system. Consider one alternative that would give REA co-op members a choice—changing the mission of the REA. The REA admirably accomplished the historic mission it was assigned in the 1930s—bringing electricity to rural America. It has now moved beyond that mission to the point where it serves as a conduit of funds from the federal treasury to electric power development in general, arranging loans for joint generating and transmission projects involving investor-owned utilities and rural electric cooperatives, and even financing coal mines.

Rechristened the Rural Energy Administration, the REA could take on an historic new mission, appropriate to the nation's needs in the 1980s. Taking advantage of its network of cooperatives that reaches into every rural community, the new REA could arrange for long-term, low-interest loans to finance a nation-wide experiment in decentralized energy. Federal monies would go into tightening rural homes and the resourcefulness of rural citizens would be channeled into developing new ideas in energy conservation and renewable resources that would benefit the entire nation.

That is an example of the kind of change required to bring about a substantial redirection of energy policy. But just imagine the resistance it would encounter. Powerful institutions and interests, including the generation and transmission cooperatives, the investor-owned utilities, and the REA itself would oppose the change. This clearly illustrates the tremendous inertia in the energy system.

One clear distinction between the hard and soft energy paths is their impact on rural Americans. Farmers and other rural people are targeted to make the major sacrifices if we continue to move along the hard energy path. Soft energy strategies, with their emphasis on conservation, tend to distribute the sacrifices much more equitably between urban and rural people.

The targeting of rural citizens is clear in the case of coal, where the most dramatic new energy developments are planned. The word from Washington is that we will move strongly into coal —coal for electricity and for synfuels (substitutes for petroleum and natural gas, derived from coal). If present plans are carried out, there will be many more strip mines, many more powerlines, many more pipelines, and many more coal trains in the rural areas of the western states.

The same is true of other centralized supply options on the hard energy path. Nuclear power is stalled now, the chief technical problems being reactor safety and radioactive waste disposal. The most likely technical fixes for both these problems will impact directly on rural people. The president's task force on Three Mile Island lists as one of its major "reforms" that future nuclear plants be built away from population centers. Just what geologic formations will be chosen for radioactive waste disposal is not yet certain, but one sure bet is that the disposal sites will not be in the cities or the suburbs. In the meantime, presently functioning reactors need fuel, and uranium companies are out combing the western states in search of promising new mine sites.

Even solar energy would target rural citizens if hard path options are pursued. Solar energy is not one thing—there are many ways the sun's rays can be harnessed. Generally speaking they fall into two major categories: decentralized (on-site) applications, such as solar collectors to heat individual homes or buildings and centralized applications, such as the "power tower" or the solar power satellite (sps) that would serve as large electric generating facilities.

There is going to be a fight over the direction of our nation's solar effort. An adequate account of the issues involved is far beyond the scope of this book, but a few comments are in order. Decentralized applications of solar are ready now and some reputable analysts estimate they could provide on the order of 20 to 25 percent of U.S. energy by the year 2000 if a large-scale effort were launched. Centralized applications are currently in the research and development stage; they cannot contribute significantly until after the turn of the century. Decentralized solar would substitute for energy now provided by energy companies; centralized solar would fit right into the electric grid.

With centralized solar, rural citizens will be called upon to make the major sacrifices. For example, the epitome of centralized solar would be the solar power satellite system, for which enthusiasm is beginning to build in Washington. As now envisioned, sps would have sixty satellites in orbit, converting sun-

POWERLINE

light to electricity, and beaming the energy down by microwaves to sixty receiving antennas on the earth's surface. With an array of photovoltaic cells several miles on a side, each satellite would generate five thousand megawatts of electrical power, equal to that of five large conventional power plants. The sixty receiving antennas would each occupy fifty-five square miles and would naturally be sited where there are "no people"—that is, somewhere out in rural America. If past experience is any guide, the particular rural citizens whose farms, ranches, and reservations are to be taken, their neighbors who may worry about the microwaves, and the other people who will be in the paths of the hundreds of high voltage powerlines that radiate from the receiving antennas, will not be identified until after billions of dollars have been invested in the project.

As the 1980s begin, America is barreling down the hard energy path. It will take a powerful countervailing force to blunt its enormous momentum and create the conditions where new directions in energy policy are possible. What the Minnesota experience suggests is that rural people will contribute a vital component of that countervailing force. What happened with the farmers and the powerline shows a depth of rural opposition and a strength of rural resistance that is far greater than policy-makers in Washington now imagine.

This resistance is not something that better public relations or technical fixes will be able to handle. Rural people everywhere are finding their lifestyles threatened by a combination of social forces. They are a minority, bucking historical trends and powerful interests who would readily sacrifice them for the "greater social good." Just as the CU powerline became a symbol of the larger threat to Minnesota farmers, other rural people can be expected to react the same way to other hard energy path projects. Witness the words of one Montana rancher that we found scribbled in Patty Kakac's notebook, likening the ranchers' fight against stripmining to the Indians' earlier fight for survival on that same land: "To those of you who would exploit us, do not underestimate the people of this area. Do not make the mistake of lumping us together as 'overburden' and dispense with us as nuisances. Land is historically the central issue in any war. We are the descendants spiritually, if not actually, of those who fought for this land once, and we are prepared to do it again. We intend to win."

The Minnesota powerline struggle demonstrates that the rural minority, fighting for survival on its own territory, has hitherto unsuspected determination and leverage. Back in Washington, up on Capitol Hill and in the agencies downtown, the traditional energy supply interests clearly have the upper hand. But out in the fields of Minnesota, in the Black Hills of South Dakota, and on the range in Montana, rural people are getting together to make a stand.

Epilogue

On April 27, 1980, Verlyn Marth died of a heart attack on his farm in Herman, Minnesota. He was fifty-six years old. Six short years before, at the "information meeting" for Grant County farmers, his fiery words had ignited a powerline protest.

At that time, Verlyn Marth was probably the only person involved who understood clearly what the powerline struggle meant and what it could become. Owen Heiberg, editor of the *Herman Review*, remembers his prophetic remark as they sat together at an early meeting: "They may get this particular line through, but that's the end. This will be the Titanic of powerlines."

Among the group gathered in the small chapel of the First Methodist Church at Herman to mourn his passing were people who spanned the protest: friends and neighbors from Grant County, including Jim Nelson and Gerry Bates, who had been at the first meeting; farmers from Pope and Stearns County, Henrietta McCrory, Mike, Anne, and Jane Fuchs, Math Woida, and John Tripp, who had been stirred by Marth's words and his vision; and younger people, Patty Kakac, Dick Hanson, Wayne Anderson, George Crocker, and Debbie Pick, who had later assumed roles of leadership in the struggle, recipients of the legacy of Verlyn Marth.

We best remember Verlyn from conversations with him at his home. He told us: "What this powerline really boils down to, I think, is a violation of the civil rights of rural people who are treated like a colony to be exploited and have no political power to do anything about it. . . . They are using the law to make farmers political prisoners on their own land.

"What it finally boils down to is this. When you have such an obvious and tragic derogation of the whole purpose of laws, which is to keep a civil, stable, society going, then you get into what we're heading for now, which is a full-scale powerline guerrilla warfare."

At the very end of the memorial service, Verlyn Marth's oldest son and his daughter rose to read some passages from a diary their father had kept. His own words express concisely the core values that underlie the farmers' passionate and furious rebellion:

A man is rich in proportion to what he leaves alone,

and,

The land the way it was;
The man the way he was.

Glossary

AC—Alternating current

BCA—Bureau of Criminal Apprehension (Minnesota's state version of the FBI)

BTU—British Thermal Unit (a unit of energy; one gallon of gasoline has an energy content of approximately 5.3 million Btu)

BURNS AND MC DONNELL—An engineering consulting firm, based in Kansas City, Missouri

THE CITIES—To Minnesotans, this means the twin cities of Minneapolis and St. Paul

COMMONWEALTH ASSOCIATES—An engineering consulting firm, based in Jackson, Michigan

CO-OP—Cooperative; used to signify either local distribution cooperatives or generation and transmission cooperatives

CPA—Cooperative Power Association, a generation and transmission cooperative based in Edina, Minnesota, which serves nineteen distribution cooperatives in western and southern Minnesota

CU PROJECT—A generic term for the lignite strip mine near Underwood, North Dakota, the Coal Creek generating station and the powerline to Minnesota (C stands for Cooperative Power Association and U for United Power Association)

CURE—Counties United for a Rural Environment, the first umbrella organization for powerline protest groups

DC—Direct current

DFL—Democratic-Farmer-Labor Party, Minnesota's version of the national Democratic Party

DNR—Minnesota Department of Natural Resources

EIS—Environmental Impact Statement

ENERGY AGENCY—Minnesota Energy Agency

EQB—Minnesota Environmental Quality Board (prior to 1977, it was known as the Environmental Quality Council)

EQC—Minnesota Environmental Quality Council, established in 1973

FACT—Families Are Concerned Too, a Pope County protest group

FALKIRK—Falkirk Mining Company, a wholly-owned subsidary of North American Coal Company, created in 1974

FFB—Federal Financing Bank

FLA—Farmer-Labor Association

GASP—General Assembly to Stop the Powerline, the umbrella protest group, from January 1978; based in the town of Lowry

G & T—Generation and Transmission Cooperative

HVDC—High-voltage, direct current

HVTL—High-voltage transmission line

IOU—Investor-owned utility

KTO—Keep Towers Out, a Stearns County protest group

KV—Kilovolt (1000 volts)

MAPP—Mid-continent Area Power Pool

MARCA—Mid-continent Area Reliability Coordination Agreement

MPIRG—Minnesota Public Interest Group

MW—Megawatt (million watts)

NORTH AMERICAN—North American Coal Company

NPL—No Powerline, a Grant County protest group

NSP—Northern States Power Company, Minnesota's largest investor-owned utility

PGC—Preserve Grant County, a Grant County protest group

POWER COMPANY—Literally, a privately owned utility; in this book, however, we follow the common usage in western Minnesota, where the power cooperatives are also frequently referred to by the farmers as power companies

REA—Rural Electrification Administration

SOC—Save Our Countryside, a Douglas County protest group

TOOPA—Towers Out of Pope Association, a Pope Country protest group

UPA—United Power Association, a generation and transmission cooperative based in Elk River, Minnesota, which serves fourteen distribution cooperatives in Northeastern Minnesota and one in Northwestern Wisconsin

USBR—United States Bureau of Reclamation

Paul Wellstone (1944–2002) was professor of political science at Carleton College in Northfield, Minnesota. He was a political organizer before being elected to the U.S. Senate in 1990 and 1996. His untimely death in a plane crash during the 2002 election galvanized public interest in his vision for progressive politics. His work, ideas, and beliefs are described in his book *The Conscience of a Liberal: Reclaiming the Compassionate Agenda,* available in paperback from the University of Minnesota Press.

Barry M. Casper is professor of physics at Carleton College. He is the author of *Lost in Washington: Finding the Way Back to Democracy in America.*

Senator Tom Harkin has represented Iowa in Congress since 1975 and has served in the U.S. Senate since 1985. He is a lifelong advocate for America's family farms and rural communities.